Space, Power and the Commons

Across the globe, political movements opposing privatisation, enclosures, and other spatial controls are coalescing towards the idea of the 'commons'. As a result, struggles over the commons and common life are now coming to the forefront of both political activism and scholarly enquiry. This book advances academic debates concerning the spatialities of the commons and draws out the diverse materialities, temporalities, and experiences of practices of commoning.

Part I, 'Materialising the commons', focuses on the performance of new geographical imaginations in spatial and material practices of commoning. Part II, 'Spaces of commoning', explores the importance of the turn from 'commons' to 'commoning', bringing together chapters focusing on the 'doing' of commons, and how spaces, materials, bodies, and abstract flows are intertwined in these complex and excessive processes. Part III, 'An expanded commons', explores the broader registers and spaces in which the concept of the commons is at stake and highlights how and where the commons can open new areas of action and research. Part IV, 'The capture of the commons', questions the particular interdependence of 'the commons' and 'enclosure' assumed within commons literature framed by the concept of neoliberalism.

Providing a comprehensive introduction to the diverse ways in which ideas of the commons are being conceptualised and enacted both throughout the social sciences and in practical action, this book foregrounds the commons as an arena for political thought and sets an agenda for future research.

Samuel Kirwan is a post-doctoral research associate, Socio-Legal Studies, University of Bristol, UK.

Leila Dawney is Senior Lecturer in Human Geography, University of Brighton, UK.

Julian Brigstocke is Lecturer in Human Geography, Cardiff University, UK.

Routledge Research in Place, Space and Politics Series
Series edited by Professor Clive Barnett
Professor of Geography and Social Theory, University of Exeter, UK.

This series offers a forum for original and innovative research that explores the changing geographies of political life. The series engages with a series of key debates about innovative political forms and addresses key concepts of political analysis such as scale, territory and public space. It brings into focus emerging interdisciplinary conversations about the spaces through which power is exercised, legitimized and contested. Titles within the series range from empirical investigations to theoretical engagements and authors comprise of scholars working in overlapping fields including political geography, political theory, development studies, political sociology, international relations and urban politics.

Published
Urban Refugees
Challenges in Protection, Service and Policy
Edited by Koichi Koizumi and Gerhard Hoffstaedter

Space, Power and the Commons
The Struggle for Alternative Futures
Edited by Samuel Kirwan, Leila Dawney, and Julian Brigstocke

Forthcoming
Nation Branding and Popular Geopolitics in the Post-Soviet Realm
By Robert A. Saunders

Political Street Art: Communication, Culture and Resistance in Latin America
By Holly Ryan

Geographies of Worth: Rethinking Spaces of Critical Theory
By Clive Barnett

"The commons has become a crucial keyword in the language of critical theory, social movements, and activism in many parts of the world. *Space, Power and the Commons* builds upon recent debates and theoretical developments to present fascinating explorations of the materiality of the commons, particularly emphasizing the relevance of "commoning": the shared praxis that composes the commons. It also makes substantial contributions to the conceptualization of the commons, shifting its definition from a focus on the moment of neoliberal 'capture' to a full appraisal of its materiality and potentialities – its *promise*."

Sandro Mezzadra, Associate Professor of Political Theory, University of Bologna, co-author of *Border as Method, or, the Multiplication of Labor* (with Brett Neilson, Duke University Press, 2013).

Space, Power and the Commons

The struggle for alternative futures

**Edited by
Samuel Kirwan, Leila Dawney
and Julian Brigstocke**

Routledge
Taylor & Francis Group

LONDON AND NEW YORK

First published 2016 by Routledge

2 Park Square, Milton Park, Abingdon, Oxfordshire OX14 4RN
52 Vanderbilt Avenue, New York, NY 10017

Routledge is an imprint of the Taylor & Francis Group, an informa business

First issued in paperback 2020

British Library Cataloguing in Publication Data
A catalogue record for this book is available from the British Library

Library of Congress Cataloging in Publication Data
Names: Kirwan, Samuel. | Dawney, Leila. | Brigstocke, Julian.Title: Space,
power and the commons : the struggle for alternative futures / edited by
Samuel Kirwan, Leila Dawney and Julian Brigstocke. Description: Milton
Park, Abingdon, Oxon : Routledge, 2016. | Series: Routledge research in
place, space and politics | Includes bibliographical references.
Subjects: LCSH: Commons—Philosophy. | Commons—Forecasting. |
Public spaces—Political aspects. | Spatial behavior—Political aspects.
| Human geography. | Human ecology. | Community life. | Alternative
lifestyles. | Power (Social sciences) | Neoliberalism--Social aspects.
Classification: LCC HD1286 .S63 2016 | DDC 333.2—dc23LC record
available at http://lccn.loc.gov/2015019500

ISBN: 978-1-138-84168-0 (hbk)
ISBN: 978-0-367-66863-1 (pbk)

Typeset in Times New Roman
by Swales & Willis Ltd, Exeter, Devon, UK

Contents

Illustrations

Figures

Table

Contributors

Dr Claire Blencowe is Associate Professor of Sociology at the University of Warwick, where she organises activities in the Authority and Political Technologies group. Claire has research interests in political subjectivity, biopolitics, vitalism, the discursive production of inequality, authority and new materialism. She is also interested in collaborative modes of thinking, writing and working. Current projects include Christianity and Citizenship in Sub-Saharan Africa and Critical Theories and Histories of the Politics of Life. Recent publications include *Biopolitical Experience: Foucault, Power and Positive Critique* (Palgrave).

Dr Patrick Bresnihan is an Irish Research Council post-doctoral scholar at Maynooth University, Ireland. He is currently completing a book that examines the (neo)liberal problematisation of scarcity and new managerial responses to the crisis of overfishing in the Irish and European fisheries. He is interested in examining the relationship between (neo)liberal thought, governmentality and enclosure in the context of worsening ecological crises, as well as the possibilities of the commons as a form of alter-biopolitics. He also participates in the Provisional University, an autonomous research project based in Dublin.

Dr Julian Brigstocke is a Lecturer in Human Geography at Cardiff University. He is the author of *The Life of the City: Space, Humour and the Experience of Truth in Fin-de-siècle* (Ashgate, 2014) and co-editor of *Authority, Experience, and the Life of Power* (Routledge, 2014). His current research focuses on sensory encounters with the future, cultural geographies of climate change, non-human attunements and the politics of aesthetics.

Dr Leila Dawney is a Senior Lecturer at the University of Brighton, UK. Her work spans social, political and cultural theory, sociology and cultural geography. Her research interests include experience, affect and embodied practice, cultures of landscape, Spinoza and new materialist theory, and the relationship between authority and community. She is currently developing research on the embodied experience of authority, and on Emergent Authorities. She has published widely on authority and experience and on politics and affect.

Dr Mara Ferreri is a Lecturer in Human Geography at the University of Durham. She completed her doctoral thesis, titled 'Occupying Vacant Spaces: Precarious Politics of Temporary Urban Reuse', in the School of Geography at Queen Mary, University of London. Among her research interests are temporary spatial practices, cultural and political action around contested urban 'regeneration' schemes and the potential for self-organisation under conditions of urban precarity.

Dr Samuel Kirwan is a Research Associate working in the Bristol University Law School on the New Sites of Legal Consciousness Project. He has a long-standing interest in the concept of the commons, having previously carried out research with community organisations and policy makers in the field of urban green space, and is currently working on how the term can enrich understandings, imaginations and practices of 'citizenship' within the sector of legal advice.

Dr Daniel Matthews is Assistant Professor of Law at the University of Hong Kong. His primary research interests are in legal theory and law and literature. His current work assesses questions of jurisdiction, drawing on resources from law, literature and continental philosophy, with a particular focus on deconstruction. He is a member of the Editorial Committee of *Law and Critique* and a regular contributor to the Critical Legal Thinking blog.

Dr Jonathan Metzger is Associate Professor of Urban and Regional Studies at the KTH Royal Institute of Technology, Stockholm. His research interests include spatial theory, the enacted ontologies and epistemologies of planning practice and more-than-human perspectives on urban planning and regional development. His recent books include *Sustainable Stockholm: Exploring Urban Sustainable Development in Europe's Greenest City* (co-edited with Amy Rader Olsson: Routledge, 2013) and *Planning Against the Political: Democratic Deficits in European Territorial Governance* (co-edited with Phil Allmendinger and Stijn Oosterlynck: Routledge, 2014).

Dr Naomi Millner is a political and cultural geographer based at Bristol University who is interested in environmental and social movements of the past and present, and how material and more-than-human presences make claims upon them. Key themes of her work and practical involvement include popular education, food and migration justice, political ecology, commons and enclosures and agro-ecology.

Professor Jenny Pickerill is a Professor of Environmental Geography at the University of Sheffield, England. Her research focuses on inspiring grass-roots solutions to environmental problems, and in hopeful and positive ways in which we can change social practices. She has published on autonomous, anarchist and indigenous activism, environmental protest camps and occupations, the emotional spaces of collective action, and online tactics.

Dr Tracey Skillington is a Lecturer in Sociology in the School of Sociology and Philosophy, University College Cork. Recent publications include: (2012) 'Climate change and the human rights challenge: Extending justice beyond the borders of the nation state', *The International Journal of Human Rights*, 16 (8); (2013) 'UN genocide commemoration, transnational scenes of mourning and the global project of learning from atrocity', *The British Journal of Sociology*; (2015) 'Climate justice without freedom: Assessing legal and political responses to climate change and forced migration', *European Journal of Social Theory*, 18 (3); and forthcoming *Climate Justice and Human Rights* (Palgrave). Her research interests include critical theory, collective memory and societal learning, social justice and democratisation.

Introduction

The promise of the commons

Leila Dawney, Samuel Kirwan and
Julian Brigstocke

> Our work is consecrated by the sun. Compared to winter days, let's say, or digging days, it's satisfying work, made all the more so by the company we keep, for on such days all the faces we know and love (as well as those I know but do not like entirely) are gathered in one space and bounded by common ditches and collective hopes. If, perhaps, we hear a barking deer nagging to be trapped and stewed, or a woodcock begging to make his hearse in a pie, we lift our heads as one and look towards the woods as one; we straighten up as one and stare at the sun, reprovingly, if it's been darkened by a cloud; our scythes and hand tools clack and chat in unison. And anything we say is heard by everyone. So there is openness and jollity.
>
> Jim Crace (2013)

A time-travelling political activist, stepping out of the upheavals of 1968, drops in to a meeting taking place in a protest camp, occupation or other gathering of the radical Left in 2015. What does she make of what she finds there? She might expect, given the gathering inequalities that constitute this global moment, a strengthened language of class, labour and resistance. She finds instead discussion similar to the above bucolic scene, one evoking a way of life constituted by common access to land and the openness of thought, relationships and life that would accompany it. She finds a language previously the domain of neo-Malthusians, back-to-the-land movements and social and legal historians: a language of the commons.

In the 1970s and 1980s, a concern for the commons was automatically a concern for the intertwined problems of limited resources, growing populations and the new forms of material poverty they engendered. It became an important area of interest in economics, anthropology and environmental science (e.g. Barrett, 1990; Dasgupta, 1982; Ehrlich and Ehrlich, 1972; Godwin and Shepard, 1979; Larson and Bromley, 1990; McCabe, 1990). The reference points for these conversations were, alongside Thomas Malthus's writing on overpopulation, Henry George's critique of private ownership of resources and land, and Garret Hardin on the tensions engendered when competing agents share access to a limited natural resource (George, 1954; Hardin, 1968; Malthus, 1973). The questions to which commons literature were oriented were those of *why* these crises were emerging, *when* this situation would reach crisis point and *how* nation-states in particular should respond to prevent it. At the risk of generalising a broad field of study, these were abstract discussions removed from the *doing* and *making* of the commons, from what it would mean for everyday relationships, practices and thinking, or oriented towards policy change and macro-economic resources management.

The commons mobilised in this text, and among the political scientists, geographers and sociologists developing the field of 'commons' studies, while not ignoring these issues, begin from a different articulation of the questions involved. If there is a similar generalisation to be made of this collection, it is that the authors seek to articulate a commons that refuses to be defined in terms of those resources in need of regulation or protection to save them from neoliberal colonisation or 'freeriders'. Rather, these commons may be spaces, experiences, resources, memories or forms of sharing and living that are positioned outside of a 'managing scarcity' agenda. The contributors have disciplinary homes throughout the social sciences and humanities, including urban planning, geography, sociology, political science and cultural theory. As such, they reflect the interdisciplinarity of emerging commons research and its ability to speak to debates across and beyond the academy.

Rather than understanding the commons as the reverse of the neoliberal market, here we analyse the commons as spatio-temporal and ethical formations that are concerned with ways of living together that resist the privatisation and individualisation of life. Nonetheless, as we consider the emergence of the language, idea and imaginaries of the commons at this present moment, it is important to recognise how they are tied to and operate through the neoliberalising forces that both restrict and produce the possibilities for common life. Increasingly, radical thinkers have understood the logics and processes of capitalist expansion, and resistance to that expansion, in terms of enclosure and commoning – a dual operation, where the possibilities for a specific form of resistance as commoning emerges through the very logic of capitalist accumulation (Midnight Notes Collective, 1990; De Angelis, 2004, 2007; Hardt and Negri, 2011; Jeffrey et al., 2012). As these forces gather pace, new enclosures in both rural and urban areas take place as capital, in its relentless move to expansion, finds more and more of the world to be subsumed into its process of commodification. These forms of power produce collective embodiment through population governance, while at the same time privatising life through neoliberal property and family relations. This generates the conditions of possibility for a turn to commoning as their counter-power.

The work of David Harvey has been central in this respect; in particular, his articulation of how late capitalism, in its expansive logic, relies on a continual process of enclosure. Capital expands by incorporating resources, people, activities and lands that hitherto were managed, organised and produced under social relations of mutual responsibility, in a process defined by Harvey as 'accumulation by dispossession' (Harvey, 2005a, 2005b). Studies across the human sciences have displayed the broad array of spaces in which the privatisation of resources through forms of enclosure is taking place. From concern about 'land grabs' in the global South enacted as responses to food insecurities in the West (e.g. Allan, 2012; Borras Jr et al., 2011; Kugelman, 2012; Magdoff, 2013; Peluso and Lund, 2011), to the dislocations and dispossessions that are part of finance capital's property bubbles (Harvey, 2011; Kappeler and Bigger, 2011; Strauss, 2009), to concerns about common access to 'gene plasma, water, indigenous knowledges, and so forth' (Swyngedouw, 2007: 52) or 'radio wavelengths, genetic diversity, knowledge of contraception, seeds, silence, streets' (*The Ecologist*, 1993: 8), these new enclosures have led to a keenly felt sense amongst those affected that the world is closing in; that there is less space to breathe (Cassier, 2006; Midnight Notes Collective, 1990; Dawson, 2010; De Angelis, 2004; May, 2013).

It is in this context that the commons has become an important resource for direct political action. As possibilities for change through traditional demo-cratic routes are seen to be dwindling,[1] with no alternative to neoliberalism as state politics being offered by any of the major parties in the English-speaking western democracies while the Troika of the European Commission, the International Monetary Fund and European Central Bank have dictated eco-nomic policy across the failing nations of the EU, activists have moved beyond the spaces of the rally, the union and the ballot box to explore different ways of doing politics. New forms of collective action emerging from anarchist DIY traditions have privileged forms of inhabitation and occupation over the form of the demonstration and rally, paving the way for the emergence of prefigura-tive spaces of resistance, most notably in the global articulations of the Occupy movement, 15-M and the Indignados. The politics of the commons takes place as a spatial response to such enclosures: from ways of negotiating collective knowledge production to the nature and politics of community, the idea of the commons provides a political idiom that evokes the collective production and claiming of conceptual and physical space. Through these emergent forms of resistance and protest, involving flat systems of management and decision making, politics without demands[2] and a focus on the making of alternative worlds, new practices of living, making and being in common are taking place as a direct response to enclosures and policies of dispossession.

The language of the commons offers above all else a new form of political language and disposition that foregrounds promise and hope above struggle and futility. It is a language that can unite active politics with environmental concerns, urban movements with rural resistance, local struggles with global politics. Its historical and cultural resonances, while at times problematic, offer resources for thinking about how to be and live otherwise; they provide imaginary architec-tures of idylls of collectivisation – spaces of possible futures where another world might take root. The idea of the commons offers a romance, and through this romance, a way forward, a way to think out of the despondent political narratives of ecological destruction, polarisation and dispossession, and a counter-narrative to that of the inevitable and uncontrollable force of neoliberalism. Above all else, it offers a glimmer of possibility that change can occur incrementally, and that small acts matter.

Yet, as the essays in this collection make clear, there is also a need to think beyond this romance, to assess where the commons emerges, the tensions it holds and what new reifications it might perform. In particular, Claire Blencowe's chapter in this volume warns of the seductive imaginaries of common life ensnaring emancipatory drives and reterritorialising such moves into relations of entrapment and capital accumulation (see also Noorani, Blencowe & Brigstocke, 2013 *Problems of Participation: Reflections on Authority, Democracy, and the Struggle for Common Life*, Lewes: ARN Press). Any understanding of the lure and promise of the idea of the commons needs to take into account the power of the melancholic genre we might call enclosure stories (McDonagh and Daniels, 2012). These have roots in the histories of the English parliamentary enclosures and in the folk practices of retelling those stories through song, story and other oral traditions as well as through literature and poetry: for example, in the poetry of JoÛ Clare; in Jim Crace's 2013 enclosure novel *Harvest*; in the folk singer Chris Wood's 2007 album *Trespasser*; in the back-to-the land movement as articulated in publications such as *The Land* and in the recent Reclaim the Fields movement active in the UK since 2011. The enclosure/commons dynamic is central in the narration of the commons by popular commons advocates such as David Bollier (Bollier, 2013, 2014). Central to this genre is the notion of loss, and an evocation of a life *before* enclosure as a narrative disrupted but ultimately left without closure, and to some extent impoverished. While culturally specific, such narratives travel and gain purchase as they are used to make links between movements and locations, and serve to bring disparate events and actions together. An attention to these different strands of the commons is central to this collection: to not only the possibilities, but also the problems of a term that combines, on the one hand, romantic imaginaries of pre-capitalist idylls, of hunter-gatherers, gleaners and primitive communities bound by their shared unity with nature and, on the other, the boundless possibilities of tecÛology, communication and human sociality.

As the essays collected in Part III of the book (An Expanded Commons) make clear, if the idea of the commons is multiplying in political and academic discourse, then what constitutes the commons is moving into new spaces too. While urban commons, creative commons and digital commons have entered everyday language, commons are also being explored in this volume through ideas of non-human participation in Jonathan Metzger's chapter, a global commons of shared memory in Tracey Skillington's chapter, the legal space of the common in Daniel Matthews' chapter and temporal commons in Julian Brigstocke's chapter. What draws many of the groups and actions described in these chapters together, alongside their resistance to dispossession, is the articulation less of a physically existing 'commons' and more of a performative claiming of the common. In this way, these new practices of political *commoning* (Brigstocke, 2014; Kanngieser, 2013) are enacting 'another world' within the neoliberal landscape, and in doing so are altering subjectivities, relations and spaces.

Indeed, since the early 2000s there has been a profound shift from thinking about 'the commons' as site or resource to thinking about commoning as a form of practice. The idea of commoning has proliferated in both intellectual and political domains as a means of thinking about collectivisation on a local scale, and as a way of valorising non-capitalist forms of social and economic organisation. The work of Peter Linebaugh has been central to this shift, and his histories of commoning and enclosure have lent powerful imagery and political traction to the invocations of the ideas of the commons discussed in this volume (Linebaugh, 2008).

In this volume, we intend to bring into dialogue some of the ways in which the idea of the commons and the turn to a politics of commoning is understood and activated. The essays move between conceptual work and empirical examples, considering the concept in law, in history and in everyday practices, in order to provide insight into the diverse ways in which this concept has gained political and theoretical purchase on the world. In particular, this collection focuses on commons and practices of commoning as they actually exist in the world, thinking about material practices of livelihood, community production, living together and getting by. Commons emerge here as objects through tecÛologies of law, political activism and everyday practice. These articulations share an ability to interrupt, or expose as contingent, a capitalist or neoliberal hegemony (Hall et al., 2013). Thus, the commons becomes a conceptual space through which alternatives are explored, experimented with and lived out, feeding into a politics of hope that not only identifies actually existing spaces of non-capitalist life but also suggests possible futures that break through the 'end of history' – the impasse of the Left in conceptualising post-capitalist futures. Indeed, Massimo de Angelis has discussed this move in terms of the 'beginning of history' (De Angelis, 2007, 2010). The idea of the commons thus raises the question of the contingency of the current economic settlement, undermining the discourse of the necessity of growth and private property. As Esteva points out, 'commoning, the commons movement, is not an alternative economy but *an alternative to the economy*' (Esteva, 2014: i149, our emphasis). Writing about, performing and thinking about the commons as an alternative to capitalist expansion, then, can be seen to participate in this world-making, and we sense in the literature around the commons a sense of its own performativity in terms of the restructuring of experience and subjectivity against the inevitability of private property and capital.

The following sections of this chapter review five fields of research on the commons. First, we explore the notion of the environmental commons, considering the way in which commons have been understood as a means of thinking about limits to environmental resources. Second, we explore more recent literature on urban commons, and on practices of spatial commoning and appropriation in urban sites. These first two fields concern the commons as a *spatial* zone of contestation and as a set of political questions concerning the occupation, management and rights to particular spaces and the opportunities that they afford. Third, we consider the idea of the commons as a *legal* concept, drawing on work in critical legal studies to understand the relationship between the commons and the law. We turn then to work in 'neo-Marxist' theory to outline how the idea of the commons has been understood in contemporary radical political thinking, before addressing finally those authors who have sought to reframe the commons debate around the concept of 'commoning'.

Environmental commons

The concept of the commons remains primarily associated with the historically important set of writings on what we might understand as 'environmental commons'. In this formulation of the concept, commons are spaces associated with what we might see as 'nature's bounty': fisheries, forests and pastures which contain resources for human exploitation and husbandry (Berkes et al., 1989; *The Ecologist*, 1993; Heynen et al., 2007; Millner, 2012; Vogler, 2000). With its roots in the ideas of the feudal commons evoked in the enclosure stories discussed above, and of the commoners' rights laid out in the English Charter of the Forest in 1217, commons are figured as natural spaces, areas of land or sea that can provide the means for human subsistence. They are understood in terms of spatially delimited natural resources that may be appropriated for use by different communities and incorporated into different economic logics. The resources contained within such environmental commons are seen to need protection, management or appropriation by different parties and are as such often sites of struggle. This approach first gained widespread attention through the work of H. S. Gordon, who argued for fisheries to be made into privately owned resources as a means of mitigating against overfishing (Gordon, 1954).

The idea of the commons as resource is perhaps most infamously articulated in Hardin's 'The Tragedy of the Commons', which maintains that commons will always have resources stripped by 'free riders'. The 'tragedy of the commons' is thus the concern that open access property regimes produce environmental crises. This perspective has been largely dismissed – with reference to etÛographic examples of commonly held property (Berkes, 1985; Berkes et al., 1989; Cox, 1985; McCabe, 1990) and through discussion of the problem of 'free riders' in economic theory and game theory (Bousquet et al., 1996; Deadman, 1999; Ostrom, 2000) – as being based on a problematic understanding of both commons and human subjects. Indeed, Hardin has also distanced himself from his earlier work (Hardin, 2007). The discourse of 'danger' around the idea of common property has nevertheless had significant ramifications. In response to these critiques, and as a way of more clearly defining these 'environmental commons', Ostrom calls such resource-spaces 'common-pool resources', rethinking the 'tragedy of the commons' through their effective management or stewardship as common pools (Dolšak and Ostrom, 2003; Ostrom, 1990; Ostrom et al., 1999).

The view of commons as resource-space is also echoed in the work of other scholars on the environmental commons who make the case for the benefits of resources managed in common (Berkes et al., 1989; Ostrom et al., 1999) rather than considering commons as a free-for-all, as was originally set out in Hardin's and Gordon's formulation. Ostrom and others have suggested that there is a difference between a 'free for all' and actually existing, governed, managed commons that are looked after and used by a bounded community who benefit as a collective. In this way, the idea of the commons as commonly managed resource can be seen to derive from a re-engagement with the social history of lands held in common and perhaps an epochal nervousness regarding 'those global resources which are not arrogated to particular individuals or states' (Douglas, 1991: 2) – a fear regarding lack of governance of property.

Work in environmental science since the mid-1990s has also moved towards a more expanded understanding of these environmental commons. This work has drawn on discourses of global citizenship and environmental awareness, as well as on the move towards a global response to climate change, to rethink the idea of the commons in terms of a global commons (Brousseau et al., 2012; Allen, 1980). These expanded understandings of spaces in common, and the question of their management, have also moved into thinking about the 'atmospheric commons' and 'space commons' (Harrison and Matson, 2001; Soroos, 1997; Vogler, 2000). Thinking about commons as spatial demarcations of natural resources, in this light, positions them as outside and prior to social relations, so that the question of their (mis)management and maintenance comes to the fore (Brousseau et al., 2012; Dolšak and Ostrom, 2003; Ostrom, 1990; Ostrom et al., 1999; Vogler, 2000).

The politics of the environmental commons, with its emphasis on nature's bounty, may draw upon Rousseau's romantic visions of union with nature, or on the trope of the greedy capitalist land-grabber and the innocent peasant. They may draw on discourses of environmentalism and conservation, on concerns about population growth and on imaginaries of noble savages. Above all, these ideas of the commons involve a particular understanding of nature as outside the realms of human activity and as resource.

While more recent work on the commons has placed the *process* of commoning at centre stage and moved away from an understanding of commons as resource space alone, there is certainly a value in this framework. In particular, both the Midnight Notes Collective and the researchers from *The Ecologist* deal with what we might see as spatial, environmental or 'earthly' commons, have been important figures in the orchestration of this new politics of the common (Midnight Notes Collective, 1990; *The Ecologist*, 1993; Jeffries, 2011). By seeking out spaces where commoning operates, they focus on 'actually existing commons' (Eizenberg, 2012), looking at initiatives round the world where people are claiming control, resisting new enclosures and regenerating the relations that produce spaces in common: 'the social or political space where things get done and where people derive a sense of belonging and have an element of control over their lives' (*The Ecologist*, 1993: 6). Such works exemplify spaces in common, spaces where commoning works, and in doing so position 'the commons' as an ethos rather than objectifying them as resources. 'What makes commons work . . . cannot easily be encoded in written or other fixed or replicable forms' (*The Ecologist*, 1993: 11). *The Ecologist*'s (1993) agenda for 'reclaiming the commons' points to the importance of community authority in the making of the commons, focusing on relations of production and reproduction rather than on understanding the commons as a fixed set of resources. As we will see below, and through some of the contributions to this volume, the performative act of writing about actually existing commons is a political strategy that displaces the inevitability and necessity of capitalist property relations.

Urban commons

In sympathy with this view of the natural or environmental commons stands the idea of city as common. Indeed, it is through urban sites that many struggles around the spatial aspects of commoning and enclosure are being played out. Historically, as Linebaugh has argued, the city has fulfilled key functions that were essential to the smooth flow of capital, such as fortification, law and sovereignty, and trade and commerce. 'As a fort, as a court, and as a port, the city has embodied in all of these functions the principle of enclosure' (Linebaugh, 2014: 25). However, Linebaugh emphasises that cities too had their commons, until these were by and large enclosed over the course of the nineteenth century. Streets, for example, were part of the urban commons (Blomley, 2004c); until they were reduced to a place of traffic and commodity flow, they were sites of sport, theatre, carnival, music and other forms of commoning. Workers, too, had commons rights: shoemakers could keep some of the leather they worked with; weavers and tailors could keep cloth remnants; dockers could collect spillings. With industrialisation, however, a dramatic process of urban enclosure occurred alongside the rural enclosures. 'Factories began to enclose handicrafts. Markets were replaced by shops. The penitentiary replaced outdoor punishments' (Linebaugh, 2014: 36).

More recently, cities have been subjected to a variety of new forms of enclosure. David Harvey argues that the process of gentrification and displacement, whereby the people who create an interesting and stimulating everyday neighbourhood life are displaced by real-estate entrepreneurs and upper-class consumers, is the 'true urban tragedy of the commons for our times' (Harvey, 2012: 78). Ed Soja, meanwhile, highlights the many ways in which public urban spaces are being physically enclosed: 'Not only are residences becoming increasingly gated, guarded and wrapped in advanced security, surveillance, and alarm systems, so too are many other activities, land uses, and everyday objects in the urban environment, from shopping malls and libraries to razor-wire-protected refuse bins and spiked park benches to stave off incursions of the homeless and hungry. MicrotecÛologies of social and spatial control infest everyday life and pile up to produce a tightly meshed and prisonlike geography punctuated by protective enclosures and overseen by ubiquitous watchful eyes' (Soja, 2010: 42–43). Existing urban commons are continuing to be enclosed through privatisation, fortressing, surveillance, displacement, exclusion, forms of subjectification and the hegemony of capitalist spatial imaginaries (Hodkinson, 2012).

Yet the city also provides enormous potential for the production of new practices of commoning. Indeed, Hardt and Negri argue that the city itself can be seen as the ultimate contemporary common (see also Chatterton, 2010: 627). As global urbanisation continues at frantic pace, now 'the city is to the multitude what the factory was to the industrial working class' (Hardt and Negri, 2009: 250). Now that the whole city is a potential site of economic production, the spaces for resistance and for inventing alternatives to the existing order have multiplied. Some of the most prominent examples of urban commoning in recent years have come from the Occupy protest camps in Zucotti Park, Gezi Park, the space surrounding

St Paul's Cathedral and dozens of other cities around the world, including very active protests, at the time of writing, in Admiralty in Hong Kong and Parliament Square in London (Kuymulu, 2013). The protests, which originally responded to the austerity drive associated with the fall-out of the 2008 financial crash, draw inspiration from movements such as the 2010 'Arab Spring' and the Indignados in Spain and Portugal.

As Vasudevan observes in his autonomist reading of Occupy, which draws out its links to earlier occupation movements such as squatting, the term 'occupy' is used not in its more conventional sense of military conquest, but to name the process of '*building* the necessary conditions for social justice and new autonomous forms of collective life' by creating new infrastructures for the invention of new spatial practices (Vasudevan, 2015: 318). The Occupy movement materialises the social order it wants to enact, constructing an architecture of protest that generates new ways of assembling, debating, decision making, living, educating and dwelling. 'Occupation as a radical *politics of infrastructure* thus revisions the city as a set of relations that take form as alternative *common* spaces for political action' (Vasudevan, 2014: 3). Urban commoning, however, is an everyday practice that is not always, and not necessarily, ideologically opposed to state capitalism. Bresnihan and Byrne's study of independent spaces in Dublin, for example, shows that spaces that are not characterised by ideological or countercultural identity can nevertheless mobilise a powerful everyday politics of owning, producing and organising in common. 'Buildings and spaces are liberated from their simple existence as a disused office building, empty yard, pavement, or whatever. Physical space is socialized into a space of multiple uses: for living, eating, learning, listening to music etc. . . . the urban commons described here integrate people, physical space, materials, tecÛologies and knowledge' (Bresnihan and Byrne, 2014: 11). Commons are materialised through everyday practices that respond to multiple wants and needs, and which are negotiated and decided upon collectively. The common enacts a political imaginary which can be 'anti (against), despite (in) and post (beyond) capitalist' (Chatterton et al., 2012). It always, however, requires a transformation of everyday life and the spaces in which it is lived. It demands, for example, forms of living that challenge the alienating separation of production from consumption and fully recognise the conditions under which what we eat, wear or work with have been produced (Mies and Bennholdt-Thomsen, 2001). A 'commonist' transformation of the everyday would thus require a reinvention of the space of the home, so that the highly labour-intensive work of reproducing human beings can be collectivised and practised in common (Federici, 2011, 2014).

Legal commons

A rather different focus on the forms of urban protest described above has emerged from the third major area of commons studies we wish to present here, which comes from the broadly defined field of critical legal studies. Here, law plays a doubled role: both site of emancipation and critique, a sentiment pithily set out in the manifesto for the Teatro Valle collective: '[w]e use law when it proves useful. We break the law when it prevents the realization of a more just common life' (Teatro valle Occupato, translated and cited in de Lucia, 2013). As Antonia Layard has noted, while discussion of Occupy London focused upon ideology and new forms of protest, the battle over the establishment and maintenance of the camp was primarily a legal one (2012), finishing with the High Court recognising the right of the City of London to attain an order of possession, despite the land's being recognised as 'public', thus rejecting the claims of the protesters to be acting in line with common access to the land and to be necessarily defying the law in the interests of the global commons (*London* v. *Samede*).

If there is one conceptual area within law which is seen to prevent the attainment of a more just common life as described by the Teatro Valle collective, and to which legal theories of the commons have been directed, it is that of private property. Nicholas Blomley's (2003, 2004a, 2004c, 2007, 2008) work has laid bare the extent to which property law not only conditions our relationships to the things we 'own' but structures the everyday. Property law is intermeshed with material relationships, memories and experiences, and geographies of oppression and inequality. Drawing on the work of the historian Orlando Collinson, David Graeber's *Debt: The first 5,000 years* (2011) stresses that the idea of owning a thing, and being able to do as one pleases with it, would hold together as a juridical concept only if that thing were also a person – in other words, a slave (Graeber, 2011). The conception of private property, the argument runs, as derived from Roman law, is inextricably linked to the Roman practice of keeping slaves. Furthermore, central to contemporary struggles for the commons has been an engagement with an imaginary idea of a commons lived, experienced and produced under pre-enclosure subsistence arrangements. By bringing to light the particular interests that drove the dissolution of the commons and practices of commoning, as such undermining the narrative of progress that had otherwise framed this shift, such studies have laid bare the historical contingency of this process. The force of commons research has been to show that private property is not, in other words, a necessary component of a modern, functioning society but, rather, a contingent product of a particular historical process.

Despite this, property law, and academic discussion around it, persists in assuming the correspondence of private property and private space (Layard, 2012). They remain guided by the assumption that the ownership of land describes a heroic relationship between the owner and her field, home or shopping centre in which she is free to do with it as she pleases, irrespective of the practices and experiences that surround and, as recent research in performative geographies would stress, constitute it (Glass and Rose-Redwood, 2014; Sullivan, 2011). Indeed, as Jonathan Mitchell argues, if private property can be defined at all, it is through the othering of these practices; by the creation of constitutive outsiders (Mitchell, 2008). This recognition of the 'constitutive outside' of private property allows a critical perspective on how the concern for environmental commons, as scarce resources, can tend towards a call for the responsible enclosure of such commons as a more rational and productive use of a limited resource.

For the authors presented here, this notion of scarcity exists only as long as the practices that constitute fields, homes and shopping centres are subordinated to the rights of landowners. Where the law of private property assumes a space to be fixed and finished, and adjudicates on the basis of claims of title, a law that begins from the making of the commons would assume a space to be continually in the making, adjudicating on the shifting basis of the productivity of practices that constitute a land.

Such thoughts are not as far fetched as they may appear. As Daniel Matthews' chapter asserts, the common law, inasmuch as it is oriented to 'the ongoing relationships that constitute lives lived in common', retains the capacity to hold for a community that is shifting and unstable. The chapter argues for the importance of rejecting the language of the liberal rights holder and the imagination of law as a state-centred institution. A law that has become decoupled from the commons it serves, that adjudicates on the basis of private concerns or on the whims of markets, is law become tecÛocratic domination (Mitchell, 2008). As de Lucia also argues, the task becomes the regrounding of law 'in the actions and practices of bodies and communities': the practised reorientation of law towards the commons (de Lucia, 2013). Examples of such a regrounding of law in common practices – a legal commoning – are multiple. We could look to the use by unions of employment law to generate knowledge of casual contracts and employment rights, or the ways in which advocacy groups have drawn upon the human rights to a private and family life to contest decisions made in national courts based upon embedded prejudice against minority groups. In sum, while the role of law as a tool of tecÛocratic domination is without question, the law, once placed within the desires and dreams of the commons, can be used to disrupt this domination. The law is not simply *useful*, but productive and transformative as a space for struggle.

The Marxist commons

The fourth major contribution to literature on the commons has emerged from a Marxist tradition of political thought, in which the singular 'common' is reimagined in terms of human potential, enabling new forms of collective endeavour, but at the same time providing increased possibilities for capitalist accumulation. Michael Hardt's distinction between commons and common is useful here, where 'the common' is understood in terms of the social in its productive and creative force, drawing its reference points from the 'common good' and 'common ownership' as they are discussed in Rousseau and Marx. 'The commons' pertains to questions of land and resource use: to the natural world and its use by society in a similar vein to the writers concerned for the environmental commons discussed above. Thus, the common is associated with the infinite potential of human sociality. In the work of Hardt and Negri, it pertains to 'those results of social production that are necessary for social interaction and further production, such as knowledges, languages, codes, information, affects and so forth', whilst describing also the 'practices of care and cohabitation in a common world, promoting the beneficial and limiting the detrimental forms of the common'. The common thus becomes everything that could be shared or enables sharing, and the issue then becomes its 'maintenance, production and distribution' – in other words, the relations of production and reproduction (Hardt and Negri, 2011: viii).

Within this discussion is the implication that neoliberal political economy requires the privatisation, and as such the rendering available for exploitation, of both the natural commons and the 'artificial' common of human creativity and cognitive production. In terms of the latter, the work of Carlo Vercellone (2007) has been instrumental in terms of understanding how the nature of capitalist production (in the West) has shifted towards the channelling and harnessing of ideas, thoughts, desires and affects in his mapping of the contours of 'cognitive capitalism'. Vercellone's writing captures what is so troubling about contemporary capital: that it is our own capacity to creatively *disrupt* the normal that produces new spaces for capitalist expansion, as evidenced in the concept of 'disruptive innovation' as small-scale, bottom-up innovation that supersedes more established forms of business practice (Christensen, 1997).

Authors within this 'neo-Marxist' tradition have discussed how the common and practices of commoning have emerged from and through the relations of capitalism; the relations of innovation and connectivity that have emerged through the communicative expansion of capital, in the work of Hardt and Negri, and perhaps more so in Dyer-Witheford, provide the infrastructural architecture for a post-capitalist common world (Dyer-Witheford, 2006; Hardt and Negri, 2011). In these tecÛo-communisms, the conditions through which the politics of the common are emerging are produced *through* neoliberalism and can perhaps exist in harmony with neoliberalism: 'electronic media display contrary tendencies that radically subvert the logic of the market' (Dyer-Witheford, 1999: 202). Dyer-Witheford argues that the practices of hackers, peer-to-peer communities and open source programmers all 'constitute a clandestine shadow world that obstinately follows the attempt to enclose information in commodity form' (Dyer-Witheford, 1999: 439), and, following Dorothy Kidd and others, proposes the formation of a 'communications commons' as a means of achieving economic and social transformation as part of a post-work economy.

Both Hardt and Negri's and Dyer-Witheford's arguments are based on the productive, creative capacities of common life. In capitalism, the capacities of markets to seize new forms of common life and subsume them into market relations are possible only because of the possibilities that collective human life offers for the production of commons. The proliferation of human creativity and interaction through such networks provides a form of the common that is without limit, or whose limits are not material, but circumscribed by contingent forces of enclosure which may and can be resisted.

To return to a point made above with regard to legal commons, Dyer-Witheford's vision of common life is that of a commons unimpeded by scarcity; the limits of a resource-based common are here replaced by a commons of potentiality. Hardt and Negri's politics of the common makes a similar distinction between the 'common wealth of the material world' – a commons with limits – and a common whose limitlessness emerges from the forms of tecÛological reproducibility: commons that have emerged in their current form through the creative capacities of capitalist production. As capitalism moves towards economies of immaterial labour, so knowledge, information, affects and relationships can be expropriated by capital to generate surplus value. As the basis for this expropriation, the common is understood as the 'product of labour and the means of its future production . . . this common is not only the earth we share but the languages we create, the social practices we establish, the modes of sociality that define our relationships, and so forth' (Hardt and Negri, 2011: 139). In other words, the commons as defined here, as collective production that is not necessarily but is often currently tied to capitalism, is not dependent on a logic of scarcity.

Against the pre-capitalist utopias of commons imaginaries, Dyer-Witheford's Marxism for a networked age reaches into tecÛo-futures informed by the capitalist production of a shared corporeality – a biopolitical common that offers the potential for new forms of sociality and collective production that relies on and pushes against the capitalist forms of enclosure through which it emerges. Dyer-Witheford's account of a networked common resides in the promise of the network and the machine to produce and reproduce relations of creativity and human potential while considering those networks always under threat from appropriation and privatisation. In doing so, it looks to networked tecÛologies to realise the promise of the common while being aware of the possibility of their capture. Hardt and Negri's politics of the common looks to immanent practices of human sociality as inspiration for a new politics. The commons in Hardt and Negri is already here, but is currently being managed under capitalism. The multitude emerges, thus, as a 'many headed hydra' (Linebaugh and Rediker, 2000), as the poor of the world, leading to an understanding of politics as a biopolitical struggle to produce new truth and reality: 'truth . . . constructed from below . . . forged through resistances and practices of the common' (Hardt and Negri, 2011: 121). Their politics of the common are produced collectively as a process of reorientation of power away from those that dominate, pointing to a project that focuses on 'practices of care and cohabitation in a common world, promoting the beneficial and limiting the detrimental forms of the common'. The common, then, comprises everything that could be shared or enables sharing, and the issue then becomes its 'maintenance, production and distribution' – in other words, the relations of production and reproduction (Hardt and Negri, 2011: viii).

These tecÛo-commons and limitless understandings of the common bring to the fore the extent to which common life is both produced through capitalism and enabled by the forms of creativity and proliferation that capitalism, in its drive to growth, thrives on. In this story, as in Marx, capital contains within it the seeds of its own destruction, and yet, at the same time, we can argue that the notion of the common can make sense only through the logics and tecÛicities of late capitalism, a tension we further explore in the final section of this introduction.

An entangled commoning

One area of critique of these neo-Marxist approaches to the common is that, similar to the environmental commons work discussed above, the commons is approached as a resource to be managed according to one or other regime of economic governance. This has been argued to be problematic for two main reasons. First, it forces the debate about the commons to begin with the idea of its inherent possibility of failure: the story of the commons is always already a tragedy. This return to the Hardin tragedy obscures the fact that the problem is not the existence of commons as such but the logics of dispossession that seek to do away with them: by the focus on their availability for capture, rather than on the modes of capture themselves, they are always framed in terms of their vulnerability. While David Harvey (2005a, 2005b) to some extent resolves this by focusing on such logics rather than on commons themselves, he too falls prey to understanding commons only in terms of resources under threat from capital as accumulation by dispossession.

The second criticism returns to a point made throughout this introduction: that the neo-Marxist authors discussed above continue to position the subject of the commons – the commoner – as outside of the practices of mutual constitution through which commons are produced, rather than considering the practices as constitutive of both the space and the subject. Treating the commons as a resource offers a troubling nature/culture distinction, situating the non-human only as resources for humans to exploit and thus refusing to acknowledge the co-constitutive role of humans and non-humans in the production of the commons. In sum, we find the distinction between natural and intellectual commons, while useful for understanding what is at stake in the literature, somewhat problematic in so far as it relies upon the same reification of the natural that makes the 'tragedy of the commons' literature so problematic: the obscuring of how 'the natural' is historically enmeshed with, and inextricable from, social practices. These Marxist articulations of the commons can be seen to sideline attention to the natural *commons* in favour of a celebration of human capacity and imagination as the site of resistance to neoliberalism. While the neo-Marxist politics of the common can speak of the commons, there is little sense as to how the commons can then speak back: the world of the non-human, the material, the chthonic and the algorithmic, is rendered simply that of possibility and resource for the ultimately human acts of commoning and enclosure.

This anthropocentrism in commons thinking leads to a neglecting of narratives of intertwined practices that comprise commons. These practices are apparent not only in agriculture and husbandry but, as Peter Linebaugh argues, also in the making of the working class itself. In *The Magna Carta Manifesto*, Linebaugh discusses the American working class as constructed along 'four forces or vectors': ecological production and conservation by indigenous peoples; the 'hewers of wood and drawers of water' who prepared the land for its agrarian revolution; the factory proletarians, and finally 'those who carry out the invisible labors of reproduction' (2008: 244–245). His narrative is thus one in which the social *common* is inextricable from the natural commons; both are constituted through historically situated practices of *commoning*. Writers on the concept of commoning, from Kevin St Martin's counter-mapping of fish stocks to Gibson-Graham's etÛographies of community economies, to Nick Blomley's work on urban spacing, all evoke the commons as based upon lived entanglements between 'natural' and 'social': on commons as a *practice* (Blomley, 2004b, 2011; Gibson-Graham, 2006b, 2008; St Martin, 2009). In other words, they all assert that there is no commons preceding the practices that compose it.

Twisting a noun into a verb is more than simply allotting it a practice and a duration. As Julie Ristau argues, the idea of commoning 'brings to life the essential social element of the commons' (Ristau, 2011). To place the commons primarily as something that is *done*, rather than something that *is*, foregrounds the networks of relationships it is done with, as well as the inter-generational circuits through which this doing is learned. It posits the commons as animated and composed by a being that is relational and contingent, rather than a being that simply is. Not only is the commons active and dynamic, but it is also productive of objects, experiences, memories and lives inscribed with social particularity. To focus on commoning is to shift the commons from a particular concern for the use of resources, or with the wealth of human cognitive capacity, to the lived intertwining of body, material, experience and love that inscribes an *excess* into life.

The move towards commoning posits that commons do not pre-exist their making through practices of commoning (Bresnihan and Byrne, 2014; De Angelis, 2010; Linebaugh, 2008). Commoning makes the commons, just as farming makes the farm. As Linebaugh points out, all commons are embedded in a particular ecology, which involves practices of local husbandry; commoning is embedded in attachments, labour and experience (Linebaugh, 2008: 44). It is grounded and specific, just as capitalist economic relations emerge from a historically specific set of processes and relations. It follows, then, that an environmental commons can perhaps better be understood as a particular ecology of relation between human and non-human – to be thought of in terms of practices of sharing and sustainability, rather than as nature's bounty to be cared for or squandered. To frame the commons in terms of resources, as Bresnihan argues, is to frame the question of the commons in terms of the management of resources, rather than focusing on the immanent practices of commoning, and indeed would destroy that which holds them as common, for 'there are no commons without incessant activities of commoning, of (re)producing in common' (De Angelis, 2010: 955).

An agenda for commons research – themes and tensions

In the previous section, we have sought to provide an overview of the key areas in which the concept of the commons is being, and has been, addressed. We have sought to describe how the commons is at stake in these debates, shedding light upon the different perspectives that have animated these discussions. Part I of this volume 'materialises' these discussions with reference to empirical sites where the idea of the commons has been brought into play through practices of world making, political lobbying and through the law.

Jenny Pickerill's chapter focuses on the way in which the idea of the commons, and the notion of community, has become materialised in the design and practices of eco-villages. In discussing 'actually existing commons', she provides empirical examples of commons as both spaces and entangled practices. The chapter thus draws our attention to the spatial organisation and architecture of such villages, and the practices of their building as a means of producing both places and relations in common. Its focus on the material practices of living, making and working together foregrounds the entangled natures of actually existing commons: how spaces of common life are produced and reproduced through quotidian activities and contested imaginaries, how ideas of 'community' are negotiated through such activities and imaginaries, and how spatial design produces social, economic and emotional interdependencies and how such spaces negotiate their difference from the worlds around them.

Naomi Millner's chapter provides a nuanced history of the relationship between commons and enclosure in Victorian Britain, focusing on the particular politics of the Commons Preservation Society. Whilst remaining attentive to the nostalgic tendencies that have often led the society to be dismissed as indicative of a certain romantic paternalism, notably the belief in the civilising and curative powers of 'nature', Millner notes the 'unusual collectives' and unexpected political engagements brought together by the movement. The chapter undermines the tendency to organise this history into 'commons' and 'enclosure', noting instead how the theatrical moves to reclaim the commons indicated specific moments in which dispute was made possible. The chapter raises the potential for academic practices in the present to make a space for such moments by rendering visible, and contingent, practices of enclosure taking place across multiple registers.

In Chapter 3, Daniel Matthews turns to the legalities of the commons, focusing on the common law tradition through the eyes of post-foundational philosopher Jean-Luc Nancy, arguing for its potential as grounding of a politics of the common. Matthews' chapter explores the early modern foundations of common law as undergirded by divine authority, considering it as a legal expression of a 'natural law' of life in common that will always exist prior to the law, and that is practised through obligation and fealty. In doing so, he suggests clear connections between the post-foundational primacy of the social and early-modern thought. Through the work of Nancy, he allows the 'spirit of the common' to haunt contemporary legal articulations as a means of understanding our inescapable obligations based on an originary indebtedness to the world into which we are thrown.

A key theme running through this volume is the importance of the turn from the commons to 'commoning'. While many of the contributions share these concerns, Part II provides two explicit perspectives on this shift. Patrick Bresnihan's chapter, like Metzger's, takes a sharply critical view of the 'tragedy of the commons' and 'environmental commons' literature, doing so through an examination of one of the key spaces in which this narrative is evoked: the protection and regulation of fishing stocks. Against the narrative, supported by Ostrom's work in particular, in which stocks and their management are interpreted in terms of scarcity, Bresnihan's study of fishing practices in the west of Ireland foregrounds the complex *abundance* of the seas, and how the practices of fishermen, immanent to and entwined with the species they depend upon, extend into past and future. The chapter pays sharp attention to the practices, established through liberal rational logics of increasing efficiency, that undermine the commons. In doing so, it proposes an alternative tragedy of the commons: 'they are not visible until they disappear'.

In Chapter 5, Mara Ferreri contributes to the growing volume of literature on the emergence of a new politics of the commons and commoning as practices and ideologies associated with anarchist/squatting communities ('autonomous geographies'). Her chapter discusses these spaces in terms of the extent to which actually existing commons are able to travel and gain momentum outside of relatively closed activist circles. As in Pickerill's paper, the boundaries of the commons and the possibility of their porosity and travel into other spaces are examined. Ferreri's work highlights some important problems that emerge from the relationship between spaces of urban commoning and capitalist subjectivities, economies and social relations, focusing on the ways in which discourse and practices commonplace in anarchist groups are 'made public'. Her discussion of these tensions includes the possibility (which we discuss below) that such spaces merely make possible the more brutal forms of neoliberalisation by providing a safety net for the destitute. The chapter finishes with a call for renewed methodology for the study of commoning, expanding the remit of such studies beyond the stable interior of activist activities to the everyday discomforts, negotiations and struggles that traverse and exceed these domains.

Parts III and IV of the collection turn to two areas in which, in our view, there remain unresolved tensions and questions with regard to the idea of the commons: two areas of questioning to which, we argue, future research in these fields should respond. The first concerns an 'expanded commons', examining how the commons can enrich other areas of study and political practice. In Chapter 6, Jonathan Metzger focuses on the 'commons' at stake in urban planning theory, showing the important questionings – in particular those raised through the unsettling of established binaries – that would need to take place for the field to fully engage with the commons in its unruly and unpredictable form. Drawing on Haraway and Latour, the chapter calls for a new sensibility of planning that is open to various voices, human and non-human, local and non-local. Like Millner's article, Metzger's chapter points to the 'apparatuses of engagement' that would allow for planning processes open to the contingency of their own claims to authority and knowledge, and that would allow commons to take hold.

Julian Brigstocke's chapter, taking as its starting point the widespread calls within the Occupy movement to 'occupy the future', examines some ways in which time can be analysed and practised as a form of commons. Developing a theory of temporal commons, the chapter explores the aesthetic figures through which time and the future are represented in posters, artworks and advertisements that campaign for future justice. In particular, it analyses the figure of 'future generations' in discourses concerning the temporal commons. In contrast to attempts to represent future generations in the present, thus rendering them calculable and knowable, the chapter argues that the promise of the call to 'occupy the future' does not lie in tecÛiques for rendering the future co-present, but instead comes from an attunement to forms of 'time without me'.

Like Brigstocke's chapter, Chapter 8 also addresses the 'inter-generational commons', in this case in imaginaries of a 'global commons' in UN remembrance practices. Tracey Skillington draws on the thought of Walter Benjamin in order to conceptualise the complex temporalities of violence, remembrance and atrocity and the forms of excessive community they produce. Paying attention to the discourses of mourning and redemption that situate the global community in particular relations towards and against atrocity, Skillington argues that these forms of memorialisation do little to prevent future atrocities, and may in fact prevent a more open relation to the past and to the future. Like Samuel Kirwan's chapter, Skillington's makes a distinction between the commons and community, articulating a global commons based on ideas and histories 'in common', including a shared recognition of human cruelty, and a collective recognition of the links between past atrocities and current political intolerances, rather than the closure of an already constituted political community.

The second such area of tension concerns the *capture* of the commons. As we discussed above, the binary of commons and enclosure situates the idea of the commons as outside of capital relations, yet always at risk of capture and enclosure. It is imperative, we argue, to ask whether the commons can be extricated from the language and assumptions of neoliberalism, notably that of the necessary interdependency of commons and enclosure. The final two chapters explore further the extent to which forms of common life become prey to capture, but also how they may escape such capture.

In Chapter 9, Claire Blencowe teases out the invocations of the spirituality that, she argues, lies at the heart of many contemporary and historical formations and experiences of common life, at the same time reminding us of the necessity of producing life in common for late capitalism, and the possibility for affective and spiritual modes of being in common to be captured through capitalist forms. Her chapter, alongside many of the others in this volume, warns against the seductive power of spiritual experience and imaginaries of common life that may in fact ensnare us.

Samuel Kirwan's chapter shifts the language of commons and enclosure away from the physical enclosure of space to the enclosure of experience, arguing that the enclosure of urban spaces to form public parks in the Victorian period should not be assumed to have fully delimited the forms of shared experience that traverse them. While noting the ways in which the term 'community' can imply consensus and cohesion, leading to a policing of what can be seen and experienced, the chapter argues for an aesthetic shaping of public spaces open to a *commons* characterised by aesthetic disruption; the commons he describes is that which *escapes* capture by a common mode of experiencing the world. The chapter describes the public park as a privileged site for experiential commons within the city, noting the ways in which the uncontrolled natures of Gezi and Zucotti parks were central to the forms of political protest developed there.

We conclude this introduction by addressing these two areas for further research in more detail, seeking both to frame the papers presented in this collection and to propose spaces and questions for further research.

An expanded commons: intangible, more-than-human and temporal commons

A contemporary politics of the commons, we suggest, needs to expand the sites, spaces and temporalities of practices of commoning, just as practices of enclosure are continually finding new objects of commodification. Languages of the commons too often remain bound up, at least implicitly, with imaginaries of the commons as a territorially defined space (such as a field and the hedgerows that enclose it). Much of this language – words such as 'enclosure' and 'occupy' – makes it hard to escape such territorial spatial images. Yet many contemporary practices of commoning do not exhibit these kinds of spatio-temporalities. First, commons are increasingly intangible. Huge amounts of capital are now extracted from 'bits, bauds, and bytes of "digitalia" that include patented business models, accounting methods, pharmaceutical formulas, and gene sequences; copyright protected software, imagery, and music; trademarked jingles, logos, advertising slogans and branding strategies' (Coombe and Herman, 2004: 561–562). While valuable contemporary work is documenting struggles around such commons, we would argue that the commons also needs to be expanded in less familiar ways. The role of non-human life and agency, for example, needs greater attention within writing on the commons. Relatedly, the highly spatialised language of the commons has perhaps led to too little attention being paid to the *temporalities* of the commons: the ways in which past and future can also be enclosed and foreclosed or, conversely, practised as forms of shared commons.

First, then, a prominent site of contemporary forms of commoning is what Boyle (2009: 45) refers to as the 'intangible commons of the mind': the 'knowledge commons' or 'informational commons' which are being rapidly eroded through new intellectual property rights. Things once thought of as uncommodifiable, such as the human genome or non-human genetic material, are being commodified into intellectual property. Meanwhile, as the expansion of intellectual property is extended across a range of areas – Boyle (2009) identifies practices of enclosure such as business-method patents, the Digital Millennium Copyright Act, 'anti-dilution' rulings, the European Database Protection Directive and the lengthening of the time period before which copyrights enter the public domain – new forms of commoning emerge which insist upon the open and shared nature of intellectual and creative labour. The Open Source movement, for example, 'encourages us to understand the Web as a space where cultural creators, rather than corporations and consumers, are the principal actors in a virtuous cycle of exchange that produces an excess of value that will return to everyone' (Coombe and Herman, 2004: 569). Elsewhere, coalitions of non-governmental organisations, civil society organisations and nations of the global South are fighting to protect the genetic commons (Scharper and Cunningham, 2006). Meanwhile, the role of universities in privatising and professionalising the knowledge commons has received increased scrutiny. The specific spatialities of such struggles for intangible commons, however, need further attention. Knowledge commons do not fall within simple spatialities of place or territory, but are embedded within complex spatial relations via their entanglement with everyday practices.

A second important site for an expanded commons is 'more-than-human' agency (Whatmore, 2002). Since the early 2000s there has been a strong move towards recognising the active agency of non-human animals, plants and matter in the constitution of human social environments and publics. As Blue and Rock (2014) argue, 'without actual animal bodies (human and otherwise), publics would not and could not exist. All (animal) publics are heterogeneous publics, constituted through fleshy, embodied practices. The form these publics take cannot be known in advance, as they are continuously in process. They are emergent and performed rather than pre-formed constituencies . . . [They are] the discursive and material outcome of an assemblage of bodies, practices, and tecÛologies that are brought together by a particular issue that disrupts existing institutional mechanisms' (Blue and Rock, 2014: 515). Yet the notion of the commons can seem to presuppose a rigid, anthropocentric division between natural resources (the common) and the humans who use them (the commoner). What if we were to challenge this distinction, and recognise non-human life as also having shared claims on the commons? Martusewicz suggests that this requires an ethos of commoning based on love or Eros as 'the life-sustaining force that moves not only in the human community but also plays on us in our embodied relationship with the more than human community' (Martusewicz, 2005: 332). For Metzger (Chapter 6, this volume), a more-than-human commons requires slightly more than this: care (an ambition and ability to see and to hear needs); cosmopoliticising (thinking differently around the politics of what exists) and assuming responsibility (for the necessary exclusions and otherings that come with any commoning practice). We would add to this that more-than-human commoning requires attentiveness to the aesthetic strategies through which more-than-human commons become visible, thereby challenging the established order of 'common' visibilities and abilities (Rancière, 2006).

Finally, greater attention needs to be paid to the temporalities of the commons (see Brigstocke, Chapter 7, this volume). One of the most debilitating cultural effects of processes such as ongoing environmental degradation, rising levels of debt and the enclosure of knowledge commons is the way in which futures are increasingly experienced as predefined, foreclosed and homogeneous. Theorists such as Frederic Jameson argue that commodification is destroying our very capacity to imagine alternative futures; as prisoners of cultural and ideological closure, we cannot make sense of the social totality in which we are embedded, meaning that we are incapable of imagining anything beyond it (Jameson, 1982). Fisher (2014), similarly suggests that contemporary culture finds it impossible to imagine new cultural forms and modes of expression, leading to an empty nostalgia for past cultural styles. Lauren Berlant, meanwhile, describes the ways in which neoliberalism succeeds in attaching its subjects to the very processes that slowly wear them down, through forms of 'cruel optimism' (Berlant, 2011). Such processes of attenuated political imagination and cruel optimism in neoliberal structures of value, we argue, constitute forms of enclosure of what Bluedorn and Waller describe as the 'temporal commons' (Bluedorn, 2002; Bluedorn and Waller, 2006, see also; Lejano and Ericson, 2005). Practices of commoning, if they are to act as forms of prefigurative politics that summon new futures into being, must fight for the temporal commons that exist in memory, hope and poly-rhythmic social formations (Lefebvre, 2004).

The commons and their capture

One significant tension running through discussion of the commons concerns its relation to capital and enclosure. As discussed at the beginning of this introduction, to reveal something as a commons is always to suggest that contained within it is the spectre of its enclosure. In other words, commons are always *haunted* by this possibility, and indeed, as we described in the previous section, new forms of commoning develop as a specific response to regimes of enclosure and disposses-sion. Commons are at all times defined by and through their relation to capitalist forms of production and reproduction; it is perhaps testament to the mythologi-cal force of the enclosure narrative that commons are seen as being by their very nature under threat. Indeed, it is often enclosure's threat that provides the very ideological and affective grounding for the politics of commons – the urge to protect what is held in common.

Yet one effect of this counterposing of commons and neoliberalism is a poten-tial uncertainty around questioning the assumptions underpinning neoliberalism itself. Where discourses that celebrate the public's failure and its inability to manage remain powerful, the state is no longer looked to as a means of organis-ing production and reproduction. Coupled with a lived understanding that the state does not reflect the interests of the multitude, the politics of the commons are at risk of directly rejecting, or indirectly ignoring, the possibility that the state can provide a means through which to organise social and economic reproduc-tion. Yet neoliberalism proceeds through an aggressive, but not always success-ful, dismantling of such safety nets; Barbagello and Beuret, for example, note the stubborn resistance in the UK of the range of benefits and services, termed 'the social wage', that support the practices of life and reproduction (Barbagello and Beuret, 2012). As Ferreri makes clear in her contribution to this volume, the sorts of activities that independent spaces of commoning undertake in cities are actually providing services that have hitherto been the responsibility of the state, and can thus be seen to enable the rolling back of state welfare through the taking over of services by those who are unpaid.

It is essential, in this light, to ask whether moves to re-common are forms of organising that emerge through, and perhaps do not particularly challenge, a neoliberal society: to ask whether a politics of the commons, despite its focus on collectivisation and non-capitalist forms of exchange, can be seen to support the neoliberal project of dismantling the state as a means of providing infrastructure, welfare and other forms of social organisation. The contributions of Jenny Pickerill and Mara Ferreri to this volume (Chapters 1 and 5), while recognising certain 'failures' of the state to undertake certain roles effectively (for example, managing health and welfare, managing infrastructure), provide an indication of more nuanced and critical approaches to the commons in its relationship with history and capital, avoiding the tendency of anti-capitalist theorising to turn away from the state as a means of managing social and economic reproduction. The questions raised by Pickerill and Ferreri regarding the boundaries of commons are central to this debate; if the bureaucratic formations of the state, along with the categories of the public and the citizen, are being eroded, and if the idea of the common is replacing that of the public, then what future is there for those who, for whatever reason, do not or cannot participate in, or who are not welcome to participate in, the forms of commoning and worlding that commons theorists advocate over state-run systems? What are the obligations and responsibilities of the commoner, and how can actually existing commons incorporate those who cannot or will not contribute to their making?

The practice of writing the commons, of articulating where actually existing commons exist, is always political. Gibson-Graham's work in countering the hegemonic constructions of capitalism as ubiquitous and ever-powerful demonstrates how engaged academic praxis focusing on the making visible of areas of non-capitalist life has a central role in exposing the contingency of capital relations, and rearticulating the world as encompassing common forms of life as well as capitalist (Gibson-Graham, 2006a, 2006b, 2008). The project of making visible actually existing commons, whether temporal, resource-based or even modes of experience, is thus part of a wider project of dismantling the edifice of neoliberal capitalism whose assumed enormity of power and force serves only to depoliticise. To pay attention to the manifold forms of common life that weave through contemporary ways of living and relating, and to write and speak of them, is to bring them to visibility and thus to augment their authority as a way of world making. Equally, however, a political act of making visible has the potential to expose non-capitalist forms to those who may capitalise on their enclosure. While the project of valuing non-capitalist forms of economic, social and cultural life is a vital means through which change takes place, there is clearly an argument for leaving some stones unturned. An expanded definition of commons expands the coloniser's definition, too. The politics of such interventions perhaps needs closer examination.

The chapters in this volume indicate the many facets and tensions of emerging commons thinking and, as such, position the idea of the commons as a space of contestation, of questions to be asked and agendas to be brought forward. As a fruitful cultural resource that feeds political action, as a means of redefining the existing world and enacting new worlds, the idea of the commons holds above all else a form of promise, and it is this promise of the commons that we believe makes it such an exciting subject. The myths of communion and community structuring the quote at the beginning of this introduction are those very myths through which contemporary articulations of commons and common life are structured. While we need to exercise caution in celebrating such myths, examining the implications and political effects that these myths may engender, their power to move and inspire, to inform hope and to speak of other worlds is incontestable. As a mythology of promise, the idea of the commons is well deserving of its current attention and interest.

Notes

1 The 2003 anti-war protests may be seen as a crucial case in point here.
2 Jean-Luc Nancy provides further discussion and elaboration of this refiguring of politics (James, 2006; Nancy, 1991, 2000).

References

Allan J A, 2012, *Handbook of land and water grabs in Africa: Foreign direct investment and food and water security* (Routledge, London).

Allen R, 1980, *World conservation strategy. Living resource conservation for sustainable development* (International Union for Conservation of Nature and Natural Resources).

Barbagello C, Beuret N, 2012, 'Starting from the social wage' *The Commoner* **15** 159–184.

Barrett S, 1990, 'The problem of global environmental protection' *Oxford Review of Economic Policy* **6** 68–79.

Berkes F, 1985, 'Fishermen and "The tragedy of the commons"' *Environmental Conservation* **12** 199–206.

Berkes F, Feeny D, McCay B J, Acheson J M, 1989, 'The benefits of the commons' *Nature* **340** 91–93.

Berlant L G, 2011, *Cruel optimism* (Duke University Press, Durham, NC).

Blomley N, 2003, 'Law, property, and the geography of violence: the frontier, the survey, and the grid' *Annals of the Association of American Geographers* **93** 121–141.

Blomley N, 2004a, 'Un-real estate: Proprietary space and public gardening' *Antipode* **36** 614–641.

Blomley N, 2007, 'How to turn a beggar into a bus stop: Law, traffic and the "function of the place"' *Urban Studies* **44** 1697–1712.

Blomley N, 2008, 'Enclosure, common right and the property of the poor' *Social & Legal Studies* **17** 311–331.

Blomley N K, 2011, *Rights of passage: Sidewalks and the regulation of public flow* (Routledge, London).

Blue G, Rock M, 2014, 'Animal publics: accounting for heterogeneity in political life' *Society & Animals* **22** 503–519.

Bluedorn A C, 2002, *The human organization of time: Temporal realities and experience* (Stanford University Press, Stanford, CA).

Bluedorn A C, Waller M J, 2006, 'The stewardship of the temporal commons' *Research in Organizational Behavior* **27** 355–396.

Bollier D, 2013, *Silent theft: The private plunder of our common wealth* (Routledge, London).

Bollier D, 2014, *Think like a commoner* (New Society, Gabriola Island).

Borras Jr S M, Hall R, Scoones I, White B, Wolford W, 2011, 'Towards a better understanding of global land grabbing: an editorial introduction' *The Journal of Peasant Studies* **38** 209–216.

Bousquet F, Duthoit Y, Proton H, Lepage C, Weber J, 1996, 'Tragedy of the commons, game theory and spatial simulation of complex systems' Conference proceedings: *Ecology, society, economy. In pursuit of sustainable development, Université de Versailles-St Quentin en Yvelines, 23–25 May 1996.*

Boyle J, 2009, *The public domain, enclosing the commons of the mind* (Yale University Press, New Haven).

Bresnihan P, Byrne M, 2014, 'Escape into the city: everyday practices of commoning and the production of urban space in Dublin' *Antipode* **47** 36–54.

Brousseau E, Dedeurwaerdere T, Jouvet P-A, Willinger M, 2012, *Global environmental commons: Analytical and political challenges in building governance mechanisms* (Oxford University Press, Oxford).

Brigstocke J (2014) *The Life of the City: Space, Humour and the Experience of Truth in fin-de-siecle Montmartre*, Farnham: Ashgate.

Cassier M, 2006, 'New "enclosures" and the creation of new "common rights" in the genome and in software' *Contemporary European History* **15** 255–271.

Chatterton P, 2010, 'Seeking the urban common: furthering the debate on social justice' *City: Analysis of Urban Trends, Culture, Theory, Policy, Action* **14** 625–628.

Chatterton P, Featherstone D, Routledge P, 2012, 'Articulating climate justice in Copenhagen: antagonism, the commons, and solidarity' *Antipode* **45** 602–620.

Christensen C M, 1997, *The innovator's dilemma: When new tecÛologies cause great firms to fail* (Harvard Business School Press, Boston, MA).

Coombe RJ, Herman A, 2014, 'Rhetorical virtues: property, speech, and the commons on the world-wide web' *Anthropological Quarterly* **77** 559–574.

Cox S J B, 1985, 'No tragedy of the commons' *Environmental Ethics* **7** 49–61.

Crace, J. (2013) *Harvest* (Picador, London).

Dasgupta P, 1982, *The control of resources* (Basil Blackwell, Oxford).

Dawson A, 2010, 'Introduction: new enclosures' *New Formations* **69** 8–22.

De Angelis M, 2004, 'Separating the doing and the deed: capital and the continuous character of enclosures' *Historical Materialism* **12** 57–87.

De Angelis M, 2007, *The beginning of history* (Pluto Press, London).

De Angelis M, 2010, 'The production of commons and the "explosion" of the middle class' *Antipode* **42** 954–977.

de Lucia V, 2013, 'Law as insurgent critique: the perspective of the commons in Italy', blogpost in *Critical Legal Thinking* http://criticallegalthinking.com/2013/08/05/law-as-insurgent-critique-the-perspective-of-the-commons-in-italy/ (5 August).

Deadman P J, 1999, 'Modelling individual behaviour and group performance in an intelligent agent-based simulation of the tragedy of the commons' *Journal of Environmental Management* **56** 159–172.

Dolšak N, Ostrom E, 2003, *The commons in the new millennium: Challenges and adaptations* (The MIT Press, Cambridge, MA).

Douglas R, 1991, 'The commons and property rights: towards a synthesis of demography and ecology', in *Commons without tragedy: Protecting the environment from overpopulation – – a new approach* Ed R V Andelson (Shepheard-Walwyn, London) pp 1–26.

Dyer-Witheford N, 1999, *Cyber-Marx: Cycles and circuits of struggle in high-tecÛology capitalism* (University of Illinois Press, Urbana).

Dyer-Witheford N, 2006, 'The circulation of the common', in *Immaterial labour, multitudes and new social subjects: Class composition in cognitive capitalism*, Kings College, University of Cambridge.

The Ecologist, 1993, *Whose common future: Reclaiming the commons* (Earthscan Publications Ltd, London).

Ehrlich P R, Ehrlich A H, 1972, *Population, resources, environment: Issues in human ecology* (Freeman, San Francisco).

Eizenberg E, 2012, 'Actually existing commons: three moments of space of community gardens in New York City' *Antipode* **44** 764–782.

Esteva G, 2014, 'Commoning in the new society' *Community Development Journal* **49** i144–i159.

Federici S, 2011, 'Feminism and the politics of the commons', in *The Commoner* http://www.commoner.org.uk/wp-content/uploads/2011/01/federici-feminism-and-the-politics-of-commons.pdf.

Federici S, 2014, 'From commoning to debt: financialization, microcredit, and the changing architecture of capital accumulation' *South Atlantic Quarterly* **113** 231–244.

Fisher M, 2014, *Ghosts of my life: Writings on depression, hauntology and lost futures* (Zero, London).

George H, 1954, *Progress and poverty: An inquiry into the cause of industrial depressions and of increase of want with increase of wealth – the remedy* (Robert Schalkenbach Foundation, New York).

Gibson-Graham J K, 2006a, *The end of capitalism (as we knew it): A feminist critique of political economy* (University of Minnesota Press, Minneapolis, MN).

Gibson-Graham J K, 2006b, *A postcapitalist politics* (University of Minnesota Press, Minneapolis, MN and London).

Gibson-Graham J K, 2008, 'Diverse economies: performative practices for "other worlds"' *Progress in Human Geography* **32** 613–632.

Glass M R, Rose-Redwood R, 2014, *Performativity, politics, and the production of social space* (Routledge, New York and Abingdon).

Godwin R K, Shepard W B, 1979, 'Forcing squares, triangles and ellipses into a circular paradigm: the use of the commons dilemma in examining the allocation of common resources' *The Western Political Quarterly* **32** 265–277.

Gordon H S, 1954, 'The economic theory of a common-property resource: the fishery' *The Journal of Political Economy* 124–142.

Graeber D, 2011, *Debt: The first 5,000 years* (Melville House, New York).

Hall S, Massey D, Rustin M, 2013, 'After neoliberalism? The Kilburn manifesto' *Soundings* **53** 8–22.

Hardin G, 1968, 'The tragedy of the commons' *Science* **162** 1243–1248.

Hardin G, 2007, 'The tragedy of the unmanaged commons', in *Evolutionary perspectives on environmental problems* Ed Dustin Penn and Iver Mysterud (Aldine Transaction, Somerset) pp 105–107.

Hardt M, Negri A, 2009, *Empire* (Harvard University Press, Cambridge, MA).

Hardt M, Negri A, 2011, *Commonwealth* (Harvard University Press, Cambridge, MA).

Harrison J, Matson P, 2001, 'The atmospheric commons' *Protecting the commons* (Island Press, Washington, DC) 219–239.

Harvey D, 2005a, *A brief history of neoliberalism* (Oxford University Press, Oxford).

Harvey D, 2005b, *The new imperialism* (Oxford University Press, Oxford).

Harvey D, 2011, *The enigma of capital: And the crises of capitalism* (Oxford University Press, New York).

Harvey D, 2012, *Rebel cities: From the right to the city to the urban revolution* (Verso Books, London).

Heynen N, McCarthy J, Prudham S, Robbins P, 2007, *Neoliberal environments: False promises and unnatural consequences* (Routledge, London).

Hodkinson S, 2012, 'The new urban enclosures' *City* **16** 500–518.

James I, 2006, *The fragmentary demand: An introduction to the philosophy of Jean-Luc Nancy* (Stanford University Press, Stanford, CA).

Jameson F, 1982, 'Progress versus Utopia, or: Can we imagine the future? (Progrès contre Utopie, ou: Pouvons-nous imaginer l'avenir)' *Science Fiction Studies* **9** 147–158.

Jeffrey A, McFarlane C, Vasudevan A, 2012, 'Rethinking enclosure: space, subjectivity and the commons' *Antipode* **44** 1247–1267.

Jeffries F, 2011, 'Communication commoning amidst the new enclosures: reappropriating infrastructure' *Journal of Communication Inquiry* **35** 349–355.

Kanngieser A (2013) *Experimental Politics and the Making of Worlds*, Farnham: Ashgate.

Kappeler A, Bigger P, 2011, 'Nature, capital and neighborhoods: "Dispossession without accumulation"?' *Antipode* **43** 986–1011.

Kugelman M, 2012, *The global farms race: Land grabs, agricultural investment, and the scramble for food security* (Island Press, Washington, DC).

Kuymulu M B, 2013, 'Reclaiming the right to the city: reflections on the urban uprisings in Turkey' *City* **17** 274–278.

Larson B A, Bromley D W, 1990, 'Property rights, externalities, and resource degradation: locating the tragedy' *Journal of Development Economics* **33** 235–262.

Layard A, 2012, 'Property paradigms, place-making and localism: a right to the city – a right to the street?' *Journal of Human Rights and the Environment* **3** 254–272.

Lefebvre H, 2004, *Rhythmanalysis: Space, time and everyday life* (A&C Black, London).

Lejano R P, Ericson J E, 2005, 'Tragedy of the temporal commons: soil-bound lead and the anachronicity of risk' *Journal of Environmental Planning and Management* **48** 301–320.

Linebaugh P, 2008, *The Magna Carta manifesto: Liberties and commons for all* (University of California Press, Berkeley and Los Angeles).

Linebaugh P, 2014, *Stop, Thief! The commons, enclosures, and resistance* (Pm Press, Oakland).

Linebaugh P, Rediker M B, 2000, *The many-headed Hydra: Sailors, slaves, commoners and the hidden history of the revolutionary Atlantic* (Verso, London).

Magdoff F, 2013, 'Twenty-first-century land grabs: accumulation by agricultural dispossession' *Monthly Review* **65**(6) 1–18.

Malthus T R, 1973, *An essay on the principle of population* (Dent, London).

Martusewicz R, 2005, 'Eros in the commons: educating for eco-ethical consciousness in a poetics of place' *Ethics, Place and Environment* **8** 331–348.

May C, 2013, *The global political economy of intellectual property rights: The new enclosures?* (Routledge, London).

McCabe J T, 1990, 'Turkana pastoralism: a case against the tragedy of the commons' *Human Ecology* **18** 81–103.

McDonagh B, Daniels S, 2012, 'Enclosure stories: narratives from Northamptonshire' *Cultural Geographies* **19** 107–121.

Midnight Notes Collective, 1990, 'The new enclosures', http://www.midnightnotes.org/newenclos.html

Mies M, Bennholdt-Thomsen V, 2001, 'Defending, reclaiming, and reinventing the commons' *Canadian Journal of Development Studies* **XXII** 997–1024.

Millner N, 2012, 'New enclosures of the carbon market, Part I: Sub-Saharan "land-grabs" as accumulation by dispossession', Blog post on Antipodefoundation.org (11 January).

Mitchell J, 2008, 'What public presence? Access, commons and property rights' *Social & Legal Studies* **17** 351–367.

Nancy J-L, 1991, *The inoperative community* (University of Minnesota Press, Minneapolis, MN and London).

Nancy J-L, 2000, *Being singular plural* (Stanford University Press, Stanford, CA).

Noorani T, Blencowe C & Brigstocke J |(eds), 2013, *Problems of Participation: Reflections on Authority, Democracy, and the Struggle for Common Life*, ARN Press, Lewes.

Ostrom E, 1990, *Governing the commons: The evolution of institutions for collective action* (Cambridge University Press, Cambridge).

Ostrom E, 2000, 'Reformulating the commons' *Swiss Political Science Review* **6** 29–52.

Ostrom E, Burger J, Field C B, Norgaard R B, Policansky D, 1999, 'Revisiting the commons: local lessons, global challenges' *Science* **284** 278–282.

Peluso N L, Lund C, 2011, 'New frontiers of land control: introduction' *Journal of Peasant Studies* **38** 667–681.

Rancière J, 2006, 'Democracy, republic, representation', *Constellations* **13** 297–307.

Ristau J, 2011, 'What is commoning, anyway?', in *On the Commons*, http://www.onthe-commons.org/work/what-commoning-anyway.

Scharper SB, Cunningham H, 2006, 'The genetic commons: resisting the neoliberal enclosure of life' *Social Analysis: The International Journal of Social and Cultural Practice* **50** 195–202.

Soja E W, 2010, *Seeking spatial justice* (University of Minnesota Press, Minneapolis, MN).

Soroos M S, 1997, *The endangered atmosphere: Preserving a global commons* (University of South Carolina Press, Columbia).

St Martin K, 2009, 'Toward a cartography of the commons: constituting the political and economic possibilities of place' *Professional Geographer* **61** 493–507.

Strauss K, 2009, 'Accumulation and dispossession: lifting the veil on the subprime mortgage crisis' *Antipode* **41** 10–14.

Sullivan R, 2011, *Geography speaks: Performative aspects of geography* (Ashgate, Farnham).

Swyngedouw E, 2007, 'Dispossessing H2O: the contested terrain of water privatization', in *Neoliberal environments: False promises and unnatural consequences*. Ed N Heynem, J McCarthy, S Prudham, P Robbins (Routledge, Oxon) pp 51–62.

The Land (The Land Magazine, Bridport) 2006– Present. Print.

Vasudevan A, 2015, 'The autonomous city: towards a critical geography of occupation' *Progress in Human Geography* **39** 316–337.

Vercellone C, 2007, 'From formal subsumption to general intellect: elements for a Marxist reading of the thesis of cognitive capitalism' *Historical Materialism* **15** 13–36.

Vogler J, 2000, *The global commons: Environmental and tecÛological governance* (JoÛ Wiley, Chichester).

Whatmore S, 2002, *Hybrid geographies: Natures, cultures, spaces* (Sage, London).

Wood C, 2007, *Trespasser* (Album) R.U.F. Records.

Part I

Materialising the commons

1 Building the commons in eco-communities

Jenny Pickerill

Introduction

Eco-communities are about building and living overlapping lives. At the centre of many eco-communities is the quest to share – resources, objects, spaces, skills, and care. In eco-communities like Findhorn, Scotland (Figure 1.1), eco-homes cluster together, front doors face each other, pathways are shared, and gardens overlap. The houses seem to spill out into each other with bikes, children's toys, and plant pots filling the spaces between them.

This sharing, interaction, and mutual support represent, many eco-communities' attempts to materialise the commons – to create, build, and make space to act together. The physical and social materialities of eco-communities are purposefully built to shape the commons. The physical structures of the homes,

Figure 1.1 Eco-homes at Findhorn eco-village, Scotland.

community halls, and gardening spaces are used to encourage openness, interaction, and sharing, just as much as the social rules of a community. Too often, attention is only paid to the social materialities of the commons, to the ways in which common places are governed, maintained as open spaces, or to how shared experience is celebrated. Yet these common places to do not merely already exist; they are built, created, and redesigned, often purposefully. Eco-communities and the ways in which they plan and build their physical structures and infrastructures offer an opportunity to explore how the commons are built, what the commons are considered to be, and how the commons are intended to be used. Although the eco-communities explored in this chapter are all spatially delimited, the commons is always considered as more than just a shared place; rather, the commons is about sharing resources, objects, spaces, skills, and care. The commons in this context are about mutual support, interaction, and acting together. This interpretation of the commons emerges from a particular discourse in eco-communities around being a 'communitarian' – someone who is willing to share their life, belongings, time, and knowledge. In some eco-communities this approach goes as far as all residents communally sharing all their money.

Eco-communities are examples of actually existing commons, but they remain incomplete, partial, and sometimes problematic. Many attempt to open up their land to visitors, often creating public footpaths and welcoming signs. But while property and social relations are organised in many eco-communities to benefit all, they tend to benefit their members far more than the wider communities they are embedded in. This incompleteness, however, is not necessarily a sign of failure but, rather, an indicator of the complexity of what the commons constitute and how commoning can be practised.

Advocating community living

The term 'community' can have multiple meanings, but can be understood as 'dense, multiplex, relatively autonomous networks of social relationships. Community, thus, is not a place or simply a small-scale population aggregate, but a mode of relating' (Calhoun, 1998, 391). Community shares some ambiguity with the concepts of place and home (Willem Duyvendak, 2011). It can be understood as representing a sense of belonging (or exclusion), as a facet of identity, or a place of sharing. Indeed, Litfin suggests 'community is about moving beyond individualism to connection' (2014) while Gilman argues eco-communities value and practice 'sustainable living in human-scale community' (1991). In this chapter, the term eco-communities refers to a concern for social, economic and environmental needs and to examples of places of collaborative, collective and communal housing and living (Pepper, 1991; Chatterton, 2013a; Bird, 2010; Ergas, 2010; Kirby, 2003). Key aspirations of an eco-community include (but are not always present): a culture of self-reliance; minimal environmental impact and minimal resource use; low cost affordable approaches; extended relations of care for others (beyond the nuclear family); progressive values (for example, towards gender equality); and an emphasis on collectivist and communal sharing (Meltzer, 2005; Martin Bang, 2007; Moss Kanter, 1972; Van Schyndel Kasper, 2008). Therefore, living in eco-communities is about acknowledging the interdependency of humans with each other and nature, and practising mutual care.

In this context, eco-communities are understood to be part of the wider movement advocating *commoning*: to produce, live off and through the commons (Linebaugh, 2008). There are numerous historical examples of the commons being sustained and shared through local self-organised governance (Ostrom, 1990). In these ways common resources (such as land, forests, water etc.) are shared between nearby residents, from which all can benefit. As with earlier periods in history (such as the enclosures of the 15th and 16th centuries in England), there are continuing threats to the commons of all kinds through, for example, privatisation of water resources or patenting of indigenous knowledge. Community gardens, open source programming, reclaiming public space, and co-housing all remain on the edges of contemporary societal practices. The output from these activities has been termed peer production, which, rather than being based on monetary exchange, is valued by contributions and fulfilling needs, and based on an ethic of sharing (Blenker, 2006; Carlsson. 2008).

Living in eco-communities can require quite radical changes to life styles and economies. Many of the most established eco-communities are located in rural areas because of the space and privacy afforded them (for example, Findhorn eco-village, Scotland); however, there are also long-running urban examples (Christiania, Denmark) and an increasing focus on new urban experiments (LILAC, England, and Kailash eco-village, Los Angeles Eco-village, and Peninsular Park Commons, USA) (Table 1.1).

The term includes quite broad and diffuse examples, from intentional communities and eco-villages to co-housing. There are relatively few large (over 500 people) eco-villages; notable examples include Findhorn (Scotland), Damanhur (Italy), and Auroville (India), but most are relatively small (Conrad, 1996). Importantly, not all intentional communities or co-housing projects are eco-communities; many have no ecological imperative, and, as Sargisson (2012) suggests, in the USA co-housing may have no relation to commoning or intentional community.

In this chapter empirical material is drawn from multiple eco-communities across six countries visited in 2010: England, Scotland, Thailand, Spain, the USA, and Argentina. These case studies were chosen to reflect a diversity of eco-communities in tenure, underlying vision, build processes, and societal context. All the fieldwork was conducted by the author using a participatory action research methodological approach. The extent of participation varied between case studies. When possible I joined in activities on site such as building, gardening, scything, cooking and eating communally, engaging in group meetings, socialising, and staying on site for several days or more. Thirty-five face-to-face in-depth interviews were conducted in total. Interviews were conducted in English and Spanish. All interviewees gave written consent and were able to withdraw at any time. At each case study photographs, field diary

Table 1.1 Examples of eco-communities worldwide

Country	Name	Type
Australia	Crystal Waters	Rural
	Melliodora	Rural
	Moora Moora Co-operative	Rural
	Wolery	Rural
Canada	Whole Village	Rural
	Yarrow Ecovillage	Rural
Colombia	Atlantida Eco-village	Rural
Denmark	Christiania	*Urban*
	Dysseklide/ Torup	Rural
	Svanholm	Rural
England	BedZED	*Urban*
	Hockerton Housing Project	Rural
	Landmatters	Rural
	Tinkers Bubble	Rural
Germany	Lebensgarten	Rural
	Sieben Linden	Rural
	UfaFabrik	*Urban*
	ZEGG	Rural
India	Auroville	Rural
Israel	Kibbutz Lotan Community	Rural
Italy	Federation of Damanhur	Rural
Japan	Konohana Family	Rural
New Zealand	Earthsong Eco-Neighbourhood	Rural
Portugal	Tamera	Rural
Scotland	Findhorn eco-village	Rural
Senegal	Colufifa	Rural
Thailand	Panya Project	Rural
	Pun Pun	Rural
USA	Dancing Rabbit	Rural
	Earthhaven	Rural
	Eco-village at Ithica	Rural
	Kailash eco-village	*Urban*
	Lama Foundation	Rural
	Los Angeles eco-village	*Urban*

	Sirius	Rural
	The Farm	Rural
	Twin Oaks	Rural
Wales	Brithdir Mawr	Rural
	Centre for Alternative Technology	Rural
	Tir y Gafel	Rural

Source: Adapted from Birnbaum and Fox, 2014; Fike, 2012; Jackson and Svensson, 2002; Litfin, 2014; Miles, 2008; Miller, 2013; and Thörn, et al., 2011; Walker, 2005; White, 2002.

observations, and sketches of the site were recorded. At several sites it was also possible to access archival material.

For many proponents, such as Jackson and Svensson (2002), Leafe Christian (2003), Litfin (2014), Martin Bang (2007), the Schwarzes (1998), and Wimbush (2012), to name just a few, eco-communities are the ideal living arrangement. The organisational structure of tightly interwoven social networks and shared spaces and resources creates, in their view, the best possibilities for personal happiness, minimal environmental impact, and sustainable livelihoods. Academics such as Chatterton (2013a), Jarvis (2013), Metcalf (2004), Sargisson (2012), and Williams (2005), have also sought to critically evaluate whether the alternatives advocated in eco-communities enable more sustainable forms of living, with generally positive findings. The positive attributes of eco-community living can be broadly summarised as five overlapping conditions:

a   *Reduced environmental impact*: Sharing common infrastructures (such as energy generation, sewage systems, and water collection), sharing resources, and minimising land use through dense housing arrangements, all help communities to reduce their environmental impact. As many residents' needs are met on site (food, work, childcare, social events), the environmental impact of travel is limited. The collective mutual support for sustainable everyday practices also aids individuals' attempts to minimise their impact (Marckmann *et al.*, 2012).

b   *Increased efficiency*: Living in a community is more efficient. Resources, tasks, skills, and knowledge can all be shared (Walljasper, 2010). For example, common tasks such as child care, food production, cooking, or cleaning can benefit from economies of scale by being divided between people, rather than each person doing a little of everything (Franklin *et al.*, 2011).

c   *Socially rewarding*: There is often a strong sociality to community and for many this is more rewarding than living individually (Eraranta *et al.*, 2009). Living close to others and engaging in regular social interaction can help people meet their personal and mental needs. It facilitates mutual support and care for each other, and, for families, children can grow up with others of their own age.

d   *Self-governing*: Eco-communities are self-organised with often highly democratic or consensus-based systems for decision making. They seek to operate
autonomously from the state, often, for example, providing their own education systems, and being self-reliant in provision of housing, food, energy, and
waste disposal.

e   *Living beyond capitalism*: Finally, eco-communities operate beyond capitalist relations. Rather than generating an income by working for someone else
(in order to afford a high-consumption life style), eco-communities support
simple, often land-based, livelihoods and minimise economic needs (through
self-provision). This enables a focus on environmental and social care and creates time for more creative, innovative, and rewarding endeavours. Some eco-
communities go as far as sharing all income (Van Schyndel Kasper, 2008).

Making the commons

> We are what we live in. When we plan our buildings, we are also planning
> what kind of society we want to create . . . we make the buildings and the
> buildings make us.
>
> (Martin Bang, 2005, 124–125)

The buildings of eco-communities shape and structure many of their forms and
functions. The buildings are some of the most symbolic attributes of eco-communities, and the processes and practices of their construction and occupation
signify many of their ecological and ethical principles. In other words, the buildings could be read as representations of the intentions of eco-communities (Kraftl,
2010). Litfin (2014) argues that the physical structures of eco-villages intention-
ally shape forms of social interaction; they are what she calls 'architectures of intimacy'. These 'ecovillage landscapes have a sense of fluidity' (Litfin, 2014, 127),
illustrated by a lack of fences and the open communal space between houses, but
Litfin also asserts that a particular shape – the circle – is most conducive to sharing, equality, and communication. This is because it avoids hierarchy, everybody
can see each other, and it encourages interaction. Thus, building using circles, as
a house shape, as houses around a communal circular garden, or designing seating
in circles, encourages social interaction and may be identifiable as a particular
approach to building commons into eco-communities.

At the same time, some house-building techniques are used explicitly to
encourage community building. Seyfang (2009) argues that approaches like
straw-bale house construction can help build community because they are
inclusive, using low-cost affordable materials and enabling a broad variety of
people to get involved; 'the hand-building technique using natural materials
and little specialised labour lends itself to wider participation in building than
is the norm when specialist skills and industrial tools and materials are used'
(124–125). As a result, this method enables relationships to be built with other
people, as well as with nature through the materials used. Building collectively
not only ensures that multiple viewpoints are considered and increases a sense

of community responsibility but also, 'while working together, residents from varying social and ethnic backgrounds often find new understandings of each other and create new common ground for moving forward as supportive neighbors' (Klinker, 2004, 11). The process of building with others helps generate commonality and community.

We can use an examination of building practices, and the final buildings, of eco-communities to explore how the commons is (physically and socially) made. Such an examination brings into focus the spatialities of the commons alongside problems encountered, in other words, why these attempts at commoning place will always remain in progress, partial, and incomplete. Through this analysis it is possible to identify four key ways in which the commons are *built into* eco-communities: (1) the benefits of sharing; (2) houses are smaller, but there is more space; (3) home extends beyond the house; and (4) space for risk taking.

Benefits of sharing

The concept of economies of scale, that a certain amount of costs are fixed and if production is increased in scale, then costs reduce per unit, applies to the construction of houses and their infrastructure (Godfrey Cook, 2011). Buying construction materials in bulk often reduces the cost per house (for example, the panel construction at LILAC), and, likewise, a wind turbine can power several houses simultaneously (for example, at Hockerton Housing Project). To construct just one house would have cost proportionately more at LILAC, and one wind turbine would have produced more energy than was needed at Hockerton. In this way eco-communities benefit from economies of scale in construction.

Eco-communities are able to benefit from their size to reduce the costs of building by sharing infrastructure and devising new cost-sharing schemes. All house building requires infrastructure such as sewerage, water supply, and energy provision. The cost of this provision is obviously reduced if shared.

> Co-housing is not necessarily affordable housing. But if you look at the ongoing costs of living in co-housing, peoples' living costs are often lower because they are sharing and using fewer resources. But even if the individual homes are smaller than average, this is often balanced out by the shared costs of common interior space
>
> (Eli Spevak, interview, Peninsula Park Commons, Oregon, USA)

The Low Impact Living Affordable Community (LILAC) in Leeds, England developed a new home-ownership model to ensure the houses remained affordable in perpetuity, costs are linked to ability to pay (income), and people will not necessarily lose their homes if their circumstances change. In practice, residents pay only a housing charge (equivalent to rent, but actually purchasing equity shares) of 35 per cent of their net income. In order to make this cover the cost of the housing (acquired via a community mortgage), minimum net income levels were set for different sizes of house. This approach is possible only because it is a community project. In effect, the higher earners subsidise those on lower incomes, and yet, at the same time, they do not forfeit their investment and the approach is fair because all inhabitants pay the same percentage of their income.

Eco-communities can provide a ready pool of labour that significantly reduces costs and increases the pace of building. Labour tends to be shared in return for help in other projects, or for residents to build each house in turn, lending labour to others in return for help on one's own house.

Clinging to the steep hillside of the Sangre de Cristo Mountains north of Taos, the Lama Foundation has been building since 1968. Principally a spiritual centre – following the teachings of Ram Dass and his infamous *Be Here Now* book – it has an eclectic mixture of eco-houses. There are a large, central community dome (Figure 1.2), a log cabin, a straw-bale house, some yurts for visitors, small vault homes, a hybrid house, and many more.

Figure 1.2 Dancing in the community dome, Lama Foundation, New Mexico, USA.

The community setting encourages the building of small individual houses and the collective use of the large communal space. There are communal bathrooms, kitchen, library, music room, winter meeting room, and outdoor sheltered eating area. The whole community is off-grid, generating electricity through photovoltaic cells, using compost toilets, wood for heat, and water from an on-site spring (and some rainwater is collected). Water is heated in the main through a propane heater because solar capacity is limited. The way in which the Foundation has been set up limits residents to a maximum stay of seven years.

Building here is a collective process and part of a spiritual practice for many; one resident said they 'build with clay, mud and love'. Another noted 'building a house is so human and it has been taken away from us . . . it is so satisfying being able to build a house'. Some of the 'special places' like the stone hermitage have been built in silence and just women built others, such as the two vaults. Building at Lama is a process of sharing: sharing tools, skills, and roles (so if some people spend the day building, others will cook and provide the food); and at crucial parts of the build many people will pitch in and help:

> Building a building has to be a collective thing . . . In regular construc-
> tion it's all portioned out, you have the person who designs the build-
> ing, . . . the bulldozer people who come in and level the area . . . then you
> have the framers, then you have the insulators, then you have the dry
> wallers, then you have the painters . . . everybody is separate. . . . It's just
> so un-cohesive and it ends up costing the homeowner so much for all
> these specialised people to come in with all these really expensive spe-
> cialised tools. Whereas in natural building the same crew of people all
> build together start to finish, and you don't have to have a bunch of spe-
> cialised tools and you don't have to have a bunch of specialised knowl-
> edge. If there is someone directing you don't have to know how to use a
> nail gun or a skill saw. So it's just much more human, and then they're so
> beautiful when they're done, they just feel good.
>
> (Chelsea Lord, interview, Lama Foundation volunteer)

However, Lama has had, at times, to make compromises. These compromises have been less about saving money and more about reducing labour require-ments. As one resident noted 'you should start small and then work your way out, and so we should make sure we can cope with maintaining the buildings we currently have before we build more' (Richard Gomes, interview, Lama Foundation resident). Thus, the place is in a constant flow of moving forward and correcting earlier mistakes.

Eco-communities also benefit from scale during occupation by sharing com-mon technologies, such as washing machines, and in the density of housing. At Findhorn eco-village, Kailash eco-village, and LILAC, individual houses do not have washing machines. Instead, there is a communal laundry that saves most res-idents the financial and environmental costs of initial purchase and maintenance.

In addition, having only a communal laundry, and the slight inconvenience that introduces, can reduce how often people do their laundry. At Currumbin eco-village in Queensland, Australia, the scale of the development enabled the construction of an autonomous water management system that was not connected to mains water supplies (Tanner *et al.*, 2007).

The scale and layout of the housing will influence what other benefits can be harnessed from living in a community. Not all eco-communities have high-density housing. Often the more rural communities such as Tir y Gafel, Findhorn, and Tinkers Bubble have dispersed individual dwellings. However, those which build homes close together, especially with common walls (such as the blocks of flats at LILAC) will have a low area-to-volume ratio and low energy index. In simple terms, the fewer the external walls around, the bigger the internal space, the lower the energy required to heat the space. Thus, living closely to others reduces environmental impact in temperate regions.

Houses are smaller, but there is more space

The size of housing directly affects the resources and energy used in construction and occupation (Salomon, 2006). In the main, the smaller the units are, the more ecological they are. Most eco-communities have only small private residences because these homes need to contain only bedroom spaces (Jarvis, 2011):

> Build smaller units than normal . . . the most effective thing you can do is simply build smaller and attached housing. Most of the carbon impact of housing comes from heating it, so if you have a smaller space you do not need as much energy to heat it and if it is attached, side by side with your neighbours, then you also need less heat because the common walls share the heat across the buildings. So one of the things we do is build smaller spaces and then have common spaces to provide a little extra space.
>
> (Eli Spevak, interview, Peninsula Park Commons, Oregon, USA)

In communities such Panya Project (Thailand), there are large communal spaces for the shared kitchen, gardens, sitting area, office space, laundry, workshops, greenhouses, guest space, and bathrooms (Figure 1.3). Panya Project is near Mae Taeng, Chiang Mai, northern Thailand. Established in 2004, the ten-acre site has become a place for experimentation and education in permaculture and natural building. It was set up by a group of young Americans led by Christian Shearer with the aim of creating a permanent community. For a variety of reasons, few of the founding members stayed full time and it is now more a transient place where people come to learn skills and work the land for a few months and then move on, though several volunteers return annually. The advantage of this flux in residents, however, is that it feels quite a vibrant place, invigorated by the energy of new arrivals. All the buildings on site are described as natural buildings and the majority are earthen, built using either sun-dried adobe bricks or wattle and cob, with both techniques using clay and straw or rice husks.

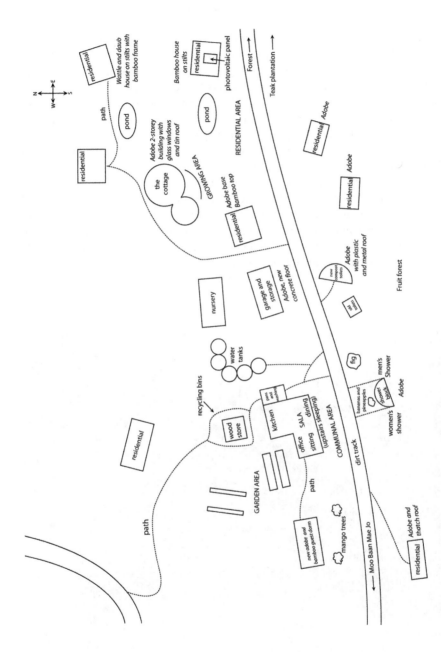

Figure 1.3 Layout of Panya Project, Chiang Mai, Thailand.

Panya tries to be about much more than the buildings, and residents consider the process of building and its completion as enhancing community (all builds are a collective process), part of a broader vision of changes required, an expression of creativity, and a nucleus of ideas that they hope people will take with them worldwide. Panya has put into practice the belief that walls and houses can isolate us from nature and each other and that if we redesign them we can better integrate nature into our daily lives. This is best exemplified in the Sala, which has few externals walls and is a very open space. It is protected from the elements by a large, over-hanging roof, but allows much of nature in. Other buildings have no glass in their windows. Many of the dwellings are also purposely small – one house has just three metres by four and a half metres of floor space. This reduces both build time and material requirements.

The residential houses need contain space only for sleeping and privacy. Most simply contain a bed and some storage space. All cooking, dining, and washing is done in communal spaces. These small houses do not need to benefit from close proximity to reduce energy use because the community is in North West Thailand and in a tropical region. Instead, the houses are built to enable air flow through them, and having some distance between structures aids natural ventilation (Figure 1.4).

Figure 1.4 A small adobe house at Panya Project, Chiang Mai, Thailand.

Home extends beyond the house

In addition to communal spaces in eco-communities enabling individual homes to be structurally quite small, the physical and emotional sense of home extends far beyond the house structures. Peninsula Park Commons is a co-housing development created by renovating some existing houses and building some new structures. Developed by Eli Spevak and Jim Labbe in 2003 in Portland, Oregon,

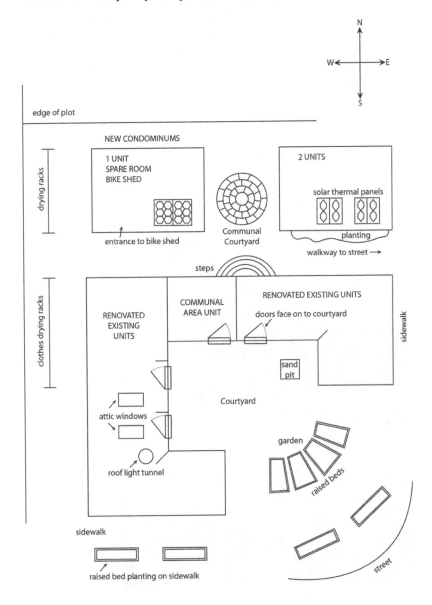

Figure 1.5 Layout of Peninsula Park Commons, Portland, Oregon, USA.

USA, it was designed to be an affordable urban eco-community. The original seven-unit courtyard apartment block was converted into six homes and a common unit (Figure 1.5). Four more new units were then built on the old driveway; 'so much of our city is already allocated to cars, so we focus on using some of that space [driveways] for homes and outdoor gathering spaces' (Eli Spevak, interview, Peninsula Park Commons, Oregon, USA). The community is ecological in its construction (reuse of building materials, solar thermal panels, use of light tunnels) and design (no car driveway but bike shed and easy bike route into the community, shared guest space, outside drying space). The common-area unit is open to all residents and has a kitchen, dining area, living room, and bathroom. It is used for watching TV, meetings, and as guest space. There is also a communal bike shed.

Figure 1.6 Peninsula Park Commons, Portland, Oregon, USA, view from the south.

Beyond the buildings are the communal gardens and raised vegetable patches (Figure 1.6). Rather than stop the development at the edge of the plot, however, Peninsula Park Commons stretched out into the street and reclaimed the pavement (sidewalk) with planters. Plant beds overflow into the placement and merge the communal garden with the public space. Nature is brought into the homes, while the community seeks to link into its neighbourhood.

As with other eco-communities, a great deal at Peninsula Park Commons is shared; in its vision statement part of its purpose has been to create 'an environment in which it is convenient to share such items as motor vehicles, home appliances, books, garden equipment, costumes, games, outdoor gear, construction tools, entertainment systems' (www.penparkcommons.org/vision.htm). There is also a very deliberate approach to existing communities; 'we want to slip into existing communities' (Eli Spevak, interview, Peninsula Park Commons, Oregon, USA). It holds community events like its annual community ice cream social and garden party.

Space for risk taking

The discussion so far has explored how eco-communities have reorganised existing places and sought to redesign these places to be more inclusive and encourage interaction. Yet eco-communities can also be supportive places for risk taking, in other words the politics of the commons encourage invention.

Lammas has built a rural eco-village (called Tir y Gafel) in west Wales. It is an unusual place in that it is one of the few eco-villages in Britain that was planned and secured planning permission before building began. Using an innovative planning policy (The Joint Unitary Development Plan (JUDP), Policy 52 – 'Low Impact Development'), Lammas was able to secure permission for nine eco-smallholdings on mixed pasture and woodland of south-facing land in Pembrokeshire. That the project was the result of years of planning appeals and allowed only through a new planning policy meant that Tir y Gafel has very much been considered at the cutting edge of eco-communities. Policy 52 set high standards for Lammas to meet: buildings were to be highly sustainable, use local, renewable, recycled, and/or natural materials, and to have low visual impact. Residents must also ensure that land-based activities (be they agriculture, forestry, or horticulture) should provide 75 per cent of basic household needs. Therefore residents had to be innovative and take risks in order to achieve all these conditions on a limited budget.

Lammas was taking a risk in pursuing the Tir y Gafel development, the first of its kind and under heavy surveillance from the state. Although residents each have their own allocated land, and thus space and freedom in how they use their plot and make their livelihood, they are collectively responsible for meeting the planning targets. One resident described it as 'more a village than a community', in that they live side by side but do not predetermine how everyone should live; yet if just one of the households fails to comply to Policy 52 the whole development is at risk of facing demolition. There was also a belief that, individually, they would struggle to build and live off the land – that without the physical and emotional support of each other it would be a long, hard task for survival. There was a strong sense of mutual solidarity, sharing, and kindness. For Lammas it was only through collectively supporting each other, being an eco-community, that they were able to take the risk to build a new eco-village from scratch.

The freedom that they have collectively has created the space for innovative and inventive eco-building. Simon Dale and Jasmine Saville had previously built an eco-house, colloquially called the 'hobbit house', but had been unable to secure long-term rights to the land on which it was built. At Tir y Gafel they have been able to create a larger version as their home and continue to experiment in a variety of natural building techniques (Figure 1.7). As Seyfang and Smith (2007) have argued, places such as Tir y Gafel provide invaluable spaces for experimentation and grassroots innovations that can develop without competition (in niches). Once completed and tested in this protected environment, these innovations can serve as models for broader-scale sustainable practices.

The risk taken by Tir y Gafel residents went beyond establishing a new community on a Welsh hillside, to challenging building regulations. Residents took a risk in not complying with standard building regulations, arguing that they were costly and not applicable to their novel eco-constructions. There was a concern

Figure 1.7 Simon Dale in his house, Tir y Gafel, Wales.

Photo by Amanda Jackson

that trying to make houses built from natural materials comply with building regulations, such as airtightness tests, would be extremely expensive and cost money the community could not afford. Complying with regulations also means paying to have structural checks and to get buildings certified by various professionals, a use of expertise that is rejected by many eco-communities. Unfortunately, the state insisted that all homes on the site should comply with current building regulations. There was a stand-off: court cases, criminal proceedings, and several homes were threatened with demolition. Many of the features of the houses at Tiy y Gafel, like the use of external compost toilets or the lack of piped water, were deemed inappropriate by the state (Dale and Saville, 2011). The result was significant remedial work and increased costs:

> Our home costs approximately £3,000 to construct. Depending upon the flexibility of building control officers, compliance would increase this cost by an estimated 100–1000% which would use up at least all our budget for establishing our land based businesses and quite likely possibly render the project unviable.
>
> (Dale and Saville, 2011, 11)

Tir y Gafel was experimenting in radical eco-house design and building methods, a vital space in which niche ideas can be tested and developed, but, unfortunately, the state sought to close down such experimentation.

Diversifying the commons

Lydia Doleman (2011), an American self-builder, argues that 'buildings have the capacity to equalize people or segregate them'. In other words, eco-communities can build places that encourage diversity or constrain it. Eco-communities need to design their buildings and shared spaces to accommodate diversity, building for different bodily abilities (Imrie, 2013). Eco-communities tend to have a more open sense of what constitutes family, rejecting the heteronormative concept of male and female coupledom that dominates in other parts of society (particularly Europe) (Wilkinson, 2013; Wilkinson and Bell, 2012). Moving beyond the nuclear family and single-family dwelling as the defining form of social structure has enabled eco-communities to develop new forms of interpersonal relationships and intimacy (a new politics of self) (Eraranta *et al.*, 2009). This is expressed through a greater acceptance of different sexualities, multiple partners and shared child care. Gender equality was also aspired to by trying to avoid creating gender-specific roles (such as men doing the building and women the cooking), and by sharing the domestic housework and child-care burden (Vestbro and Horelli, 2012). There has been mixed success in achieving this, however, as Metcalf acknowledges that 'within most intentional communities, however, we find traditional gender roles being followed by women and men' (Metcalf, 2004, 100).

This open sense of family extends to a concept which Critchlow Rodman calls 'co-care', a form of neighbourly mutual assistance that is being developed in co-housing designed particularly for seniors and the older generations (Critchlow Rodman, 2013). This approach, being practised at Wolf Willow and Harbourside co-housing in Canada and in the Netherlands, designs houses and builds community around the needs of an ageing population (Brenton, 2011). In addition to the structural provision of a carers' suite on site and disabled-accessible rooms all on one level, a sense of responsibility to look after and care for each other is built into the social elements of the community. Despite a growing recognition of the need to develop intergenerational eco-communities, issues of ageing were largely ignored in the case study eco-communities (Bauman *et al.*, 2007; Glass, 2012; Brenton, 2013). There can be high turnover in eco-communities and, as Manzella notes, 'there is little about contemporary intentional communities that encourage future generations to stay' (2010, 171); all too often children leave and communities age without an influx of younger newcomers. Even in those with multiple generations, like Findhorn eco-village, there was concern that there were no pensions or provisions planned for the long-term residents who have been members and have worked in the community for decades.

Other forms of diversity, such as race, disability, or class have been even more neglected by eco-communities (Chitewere, 2006; Chitewere and Taylor, 2010; Van Schyndel Kasper, 2008). White, middle-class, university-educated professionals have long dominated eco-communities (Pepper, 1991; Manzella, 2010). Some communities have sought to specifically engage with those who have disabilities or special needs. Inspired by Rudolf Steiner, Sólheimar (Iceland) and Kitezh (Russia) eco-communities have sought to create welcoming, supportive places for special-needs children (Dawson, 2006). Camphill communities (Britain and Norway) also seek to provide places for disabled people to live and work. A few communities have also sought to be multi-racial. Koinonia Farm (Georgia, USA), a Christian farm eco-community, explicitly sought to attract black participants, but different emphasis on materialism, social justice, and socio-economic conditions meant that 'it was difficult for blacks who earnt a wage, even a low wage, to give that up and move to a community in which communal sharing was the rule, which would amount to "voluntary poverty"' (Manzella, 2010, 173).

In the main, however, this approach creates communities for diverse others as *separate from* other eco-communities, rather than seeking to diversify residents per se. Rather, there is a significant risk of homogeneity where 'communities defined in terms of a shared home inevitably produce insiders and outsiders' (Willem Duyvendak, 2011, 83) in the ways in which boundaries are created (Moss Kanter, 1972). This may give community residents identity and power, but is problematic for others and for diversity. Indeed, 'the more diverse and powerful individuals are, the more stable and lively the community will become through a network of complex relationships . . . unity and diversity need each other' (Halbach, 2005, 28). While the value of diversity is often acknowledged, the purpose of many eco-communities is, in part, to create boundaries between their community project and mainstream society. These boundaries create space for experimentation, alternative ways of doing and living, and facilitate important feelings of belonging and identity. In this context Sargisson explores the purposeful *estrangement* eco-communities create which 'facilitates critical distance and group coherence' (Sargisson, 2007, 393). The experience of estrangement, however, is paradoxical for many eco-community members, who both need it so as to feel part of the community, but also eventually find it too much to endure. Estrangement in eco-communities acts to create members as a permanent otherness, separate, alienated, and distanced from mainstream society. To overcome the possibility of this alienation's developing into difficult practices (motivated by fear and mistrust), Sargisson argues that 'the boundaries that surround intentional communities need to be punctured and kept porous' (Sargisson, 2007, 417).

Eco-living requires time, patience, and compromise

> We are building houses, that's the easy bit, but we are also building a community. Anyone can build houses, but it's really difficult to build communities . . . we put a lot of effort into that, how we make decisions, how we have fun together.
>
> (Chatterton, 2013b)

Eco-living requires significant negotiation, compromise, and careful attention to decision-making structures (governance). Living in eco-communities is not always easy, exemplified by the sometimes high turnover of residents who find communal life too difficult (Garden, 2006). There is significant diversity in the ways in which eco-communities have sought to make decisions, from highly democratic consensus-based models to decision-making power being concentrated in a few leaders (Metcalf, 2004). The more democratic and shared the approach, the longer the process takes. The consensus model, in particular, is lengthy because decisions are reached only when there is unanimous agreement (without voting). The process consequently requires extensive discussions, modifications to a proposal, and negotiations. This can become problematic if a decision is needed in a set time frame (Renz, 2006). However, once a decision has been agreed it can be quickly implemented because time has already been taken to negotiate problems. Crucially, practising consensus productively requires training, skill, creativity, and the ability to overcome interpersonal issues (Walljasper, 2010). As noted in the opening quote and confirmed by Cunningham and Wearing's research, building structures appears to be easier than negotiating self-organisation for the community (Cunningham and Wearing, 2013).

The need for communities to make communal decisions is juxtaposed against a fear of a lack of privacy: 'the greatest fear of many people choosing a community is that they won't have enough privacy' (Leafe Christian, 2003, 148). Jarvis argues that we are in an age of isolation, of one-person households, 'a paradox whereby yearning for connectedness coexists with neoliberal policies and cultural norms which promote self-reliance and the accumulation of private property' (2011, 561). Many eco-communities deliberately reduce privacy and instead encourage more communal and collective activities, such as eating together and, in some cases, sleeping together; 'there is a loose, inverse relationship between the degree of communalism and privacy' (Metcalf, 2004, 102). For some people, and at some times, this lack of privacy can be problematic. Litfin (2014) uses the term 'ratcheting' to describe the numerous spontaneous interactions of living in close proximity. As people move around and through the eco-community, they have many random encounters with others. People often need a balance between contact and solitude.

In terms of housing, there is a need 'to find ways to meet people's privacy needs while keeping our home sites compact and not sprawled all over the landscape' (Marsh, 2003). The tendency to seek to hide from others, to create privacy by building scattered apart, increases environmental destruction and infrastructure costs (Leafe Christian, 2003). Metcalf (2004) suggests that this lack of privacy is somewhat offset by the provision of quiet prayer or meditation spaces. Lama Foundation and Findhorn eco-village both have specific quiet spaces, with Lama Foundation also having a hermitage for silent retreat. However, most of the eco-communities visited do not have these quiet spaces. Rather, co-housing has to some extent been developed to produce more privacy while not rejecting the benefits of community and communality. Co-housing 'combines the autonomy of private dwellings with the advantages of community living' (Williams, 2005, 200), or, as Sullivan-Catlin argues, co-housing could also be conceived of as 'a cooperative neighbourhood' (2004). The co-housing model is proving so popular because it enables a balance between privacy and sharing (Lietaert, 2010). Ideally, interaction is encouraged by ensuring that front doors face each other, while privacy is created for living rooms and by careful window placement (Leafe Christian, 2003).

Conclusions

An eco-community is a place in which residents illustrate a concern for the social, economic, and environmental needs of each other and nature, and where there is collaborative, collective, and communal housing and living. Eco-communities are best understood as 'a process, and not a finished product' (Van Schyndel Kasper, 2008, 20). While many advocate that eco-communities are the best way in which to build a sustainable society, they are also problematic in their homogeneity, in their use of life-style strategies as a way in which to change the world (Fotopoulos, 2000), and in their reliance on consumption of green commodities and green technologies, which perpetuates (albeit green) capitalism (Chitewere, 2006). This is not to deny the achievements of eco-communities, but to avoid an uncritical assessment.

Eco-communities are often founded on the principle of the commons, and in many ways illustrate the success of sharing space, objects, knowledge, and time. Building collectively reduces costs and environmental waste (Urban Vestbro, 2012). Developing systems of reciprocity and sharing enables eco-communities to function and people to lead comfortable lives using fewer resources. Living in compact spaces works if there is communal space available to share, particularly access to shared green spaces. Connection is encouraged in many eco-communities through the design of spaces of conviviality and communal space where residents will regularly encounter others.

However, the benefits of sharing are complicated by the need to negotiate and compromise, and the most successful systems of sharing demonstrate clear agreements for how tools, food, and space are shared. Just as there is a need for common space, so too is there a need for retreat and private space and there is a general move toward co-housing which provides privacy for residents in small, efficient houses but still encourages communal activities and sharing (Meltzer, 1999).

Yet the commons in eco-communities is about much more than simple spatial strategy to create spaces of interaction. It is not simply about ensuring that there is as much communal space as there is private. The building of the commons and practices of commoning are multi-layered processes. They involve the shared emotional support required to take risks in trying new approaches, the listening and dialogue needed to involve diverse viewpoints and diversity into a community, and the time and patience to reach democratic decisions (Fosket and Mamo, 2009). The physical structures are crucial in enabling these more social materialities of commoning to be practised. The circular shape of the Lama Foundation community dome and the spread of Peninsula Park Commons over the pavements, for example, illustrate the bringing in and stretching out of eco-communities. These spatial strategies enable the inclusive meetings to be held in a circle, or neighbours being encouraged to join in an event. In other words, the physical structures, how eco-communities are materially built, shape how the commons work (or not) in terms of sharing resources, objects, spaces, skills, and care.

References

Bauman Power, M, Krause Eheart, B, Racine, D and Karnik, N S (2007) Aging well in an intentional intergenerational community: Meaningful relationships and purposeful engagement, *Journal of Intergenerational Relationships*, 5, 2, 7–25.

Bird, C (2010) *Local Sustainable Homes*. Transition Books, Totnes, Devon.

Birnbaum, J and Fox, L (2014) *Sustainable [R]evolution: Permaculture in Ecovillages, Urban Farms, and Communities Worldwide*. Berkeley, California: North Atlantic Books.

Blenker, Y (2006) *The Wealth of Networks: How Social Production Transforms Markets and Freedom*. London: Yale University Press.

Brenton, M (2011) Cohousing: Supportive local networks in old age, in Bunker, S, Coates, C, Field, M and How, J (eds) *Cohousing in Britain*, pp. 115–124. London: Diggers and Dreamers Publications.

Brenton, M (2013) *Senior Cohousing Communities – an alternative approach for the UK?* Joseph Rowntree Programme Paper, available at: http://www.jrf.org.uk/sites/files/jrf/senior-cohousing-communities-full.pdf.

Calhoun, C (1998) Community without propinquity revisited: Communications technology and the transformation of the urban public sphere, *Sociological Inquiry*, 68, 3, 373–397, p. 391.

Carlsson, C (2008) *Nowtopia: How Pirate Programmers, Outlaw Bicyclists, and Vacant-Lot Gardeners are Inventing the Future Today*. Oakland: AK Press.

Chatterton, P (2013a) Towards an agenda for post-carbon cities: Lessons from LILAC, the UK's first ecological, affordable, cohousing community, *International Journal for Urban and Regional Research*, 37, 5, 1654–1674.

Chatterton, P (2013b) co-founder of LILAC, quoted in Haines, S 'Strong bond' of Leeds LILAC co-housing residents, *BBC News*, 1 January.

Chitewere, T (2006) Constructing a Green Lifestyle: Consumption and Environmentalism in an Ecovillage, unpublished PhD thesis, State University of New York at Binghamton.

Chitewere, T and Taylor, D E (2010), Sustainable living and community building in Ecovillage at Ithaca: The challenges of incorporating social justice concerns into the practices of an ecological cohousing community, in Taylor, D E (ed.) *Environment and Social Justice: An International Perspective (Research in Social Problems and Public Policy, Volume 18)*, Emerald Group Publishing Limited, pp. 141–176.

Conrad, J (ed.) (1996) *Eco-villages and Sustainable Communities: Models for 21st Century Living*. Forres: Findhorn Press.

Critchlow Rodman, M (2013) Co-caring in Senior Cohousing: A Canadian Model for Social Sustainability, presented at ISCA Conference, Findhorn Eco-village, Scotland, June.

Cunningham, P A and Wearing, S L (2013) The politics of consensus: An exploration of the Cloughjordan ecovillage, Ireland, *Cosmopolitan Civil Societies: An Interdisciplinary Journal*, 5, 2, available at: http://epress.lib.uts.edu.au/journals/index.php/mcs/article/view/3283.

Dale, S and Saville, J (2011) *The Compatibility of Building Regulations with Projects under New Low Impact Development and One Planet Development Planning Policies: Critical and Urgent Problems and the Need for a Workable Solution*. Available at: http://www.simondale.net/house/Building_Regulations_and_LID.pdf.

Dawson, J (2006) *Ecovillages: New Frontiers for Sustainability*. Schumacher Briefings, Totnes, Devon: Green Books.

Doleman, L (2011) *Flying Hammer Productions*. http://theflyinghammer.com/.

Eraranta, K, Moisander, J and Pesonen, S (2009) Narratives of self and relatedness in eco-communes: Resistance against normalized individualization and the nuclear family, *European Societies*, 11, 3, 347–367.

Ergas, C (2010) A model of sustainable living: Collective identity in an urban ecovillage, *Organization and Environment*, 23, 1, 32–54.

Fike, R (ed.) (2012) *Voices from the Farm: Adventures in Community Living* (2nd edition). Summertown, Tennessee: Book Publishing Company.

Fosket, J and Mamo, L (2009) *Living Green: Communities that Sustain*. Gabriola Island, Canada: New Society Publishers.

Fotopoulos, T (2000) The limitations of life-style strategies: The ecovillage 'movement' is NOT the way toward a new democratic society, *Democracy and Nature*, 6, 2, 287–308.

Franklin, A, Newton, J, Middleton, J and Marsden, T (2011) Reconnecting skills for sustainable communities with everyday life, *Environment and Planning A*, 43, 2, 347–362.

Garden, M (2006) Leaving Utopia, *The International Journal of Inclusive Democracy*, 2, 2, available at: http://www.inclusivedemocracy.org/journal/vol2/vol2_no2_Garden.htm.

Gilman, R (1991). The eco-village challenge: The challenge of developing a community living in balanced harmony – with itself as well as nature – is tough, but attainable, *In Context*, available at: http://www.context.org/iclib/ic29/gilman1/.

Glass, A P (2012) Elder co-housing in the United States: Three case studies, *Built Environment*, 38, 3, 345–363.

Godfrey Cook, M (2011) *The Zero-Carbon House*. Marlborough, Wiltshire: The Crowood Press.

Halbach, D (2005) The rainbow power of community, in Peters, V and Stengel, M (eds) *Europtopia: Intentional Communities and Ecovillages in Europe*. Ökodorf, Sieben Linden, pp. 21–33.

Imrie, R (2013) Shared space and the post-politics of environmental change, *Urban Studies*, 50, 16, 3446–3462.

Jackson, H and Svensson, K (eds.) (2002) *Ecovillage Living: Restoring the Earth and Her People*. Totnes, Devon: Green Books.

Jarvis, H (2011) Saving space, sharing time: Integrated infrastructures of daily life in cohousing, *Environment and Planning A*, 43, 560–577.

Jarvis, H (2013) Against the 'tyranny' of single-family dwelling: Insights from Christiania at 40, *Gender, Place and Culture: Journal of Feminist Geography*, 20, 8, 939–959.

Kirby, A (2003) Redefining social and environmental relations at the ecovillage at Ithaca: A case study, *Journal of Environmental Psychology*, 23, 3, 323–332.

Klinker, S (2004) Shelter and sustainable development in Kennedy, J F (ed.) *Building Without Borders: Sustainable Construction for the Global Village*, pp. 5–30. Gabriola Island, Canada: New Society Publishers.

Kraftl, P (2010) Geographies of architecture: The multiple lives of buildings, *Geography Compass*, 4, 402–415.

Leafe Christian, D (2003) *Creating a Life Together: Practical Tools to Grow Ecovillages and Intentional Communities*. Gabriola Island, Canada: New Society Publishers.

Lietaert, M (2010) Cohousing's relevance to degrowth theories, *Journal of Cleaner Production*, 18, 576–580.

Linebaugh, P (2008) *The Magna Carta Manifesto: The Struggle to Reclaim Liberties and Commons for All*. University of California Press, Berkeley.

Litfin, K T (2014) *Eco-villages: Lessons for Sustainable Community*. Cambridge: Polity Press.

Manzella, J C (2010) *Common Purse, Uncommon Future: The Long, Strange Trip of Communes and Other Intentional Communities*. Oxford: Praeger.

Marckmann, B, Gram-Hanssen, K and Haunstrup Christensen, T (2012) Sustainable living and co-housing: Evidence from a case study of eco-villages, *Built Environment*, 38, 3, 413–429.

Marsh, C (2003) of Earthhaven eco-community, USA, quoted in Leafe Christian, D (ed.) *Creating a Life Together: Practical Tools to Grow Ecovillages and Intentional Communities*. Gabriola Island, Canada: New Society Publishers.

Martin Bang, J (2005) *Ecovillages: A Practical Guide to Sustainable Communities*. Edinburgh: Floris Books.

Martin Bang, J (2007) *Growing Eco-Communities: Practical Ways to Create Sustainability*. Edinburgh: Floris Books.

Meltzer, G (1999) Cohousing: Linking communitarianism and sustainability, *Communal Societies*, 19, 85–100.

Meltzer, G (2005) *Sustainable Community: Learning from the Cohousing Model*. Crewe: Trafford.

Metcalf, W (2004) *The Findhorn Book of Community Living*. Forres, Scotland: Findhorn Press.

Miles, M (2008) *Urban Utopias: The Built and Social Architectures of Alternative Settlements*. London: Routledge.

Miller, T (ed.) (2013) *Spiritual and Visionary Communities: Out to Save the World*. Farnham, Surrey: Ashgate.

Moss Kanter, R (1972) *Commitment and Community: Communes and Utopias in Sociological Perspective*. London: Harvard University Press.

Ostrom, E (1990) *Governing the Commons: The Evolution of Institutions for Collective Action*. New York: Cambridge University Press.

Pepper, D (1991) *Communes and the Green Vision: Counterculture, Lifestyles and the New Age*. London: The Merlin Press.

Renz, M A (2006) Pacing consensus: Enacting, challenging, and revising the consensus process in a cohousing community, *Journal of Applied Communication Research*, 34, 2, 163–190.

Salomon, S (2006) *Little House on a Small Planet*. Guilford, Connecticut: The Lyons Press.

Sargisson, L (2007) Strange places: Estrangement, utopianism, and intentional communities, *Utopian Studies*, 18, 3, 393–424, p. 393.

Sargisson, L (2012) Second-wave cohousing: A modern Utopia?, *Utopian Studies*, 23, 1, 28–56.

Schwarz, W and Schwarz, D (1998) *Living Lightly: Travels in Post-Consumer Society*. Oxfordshire: John Carpenter Publishing.

Seyfang, G (2009) *The New Economics of Sustainable Consumption: Seeds of Change*. Basingstoke: Palgrave Macmillan.

Seyfang, G and Smith, A (2007) Grassroots innovations for sustainable development: Towards a new research and policy agenda, *Environmental Politics*, 16, 4, 584–603.

Sullivan-Catlin, H (2004) 'A good borderland': Cohousing communities and social change, *Communal Societies*, 24, 121–141.

Tanner, C J, Title, B and Reedman, T (2007) A case study for the ecovillage at Currumbin – integrated water management planning, design and construction, *WIT Transactions on Ecology and the Environment*, 1, 33–41.

Thörn, H, Wasshede, C and Nilson, T (2011) *Space for Urban Alternatives? Christiania 1971–2011*. Vilnius, Balto Print.

Urban Vestbro, D (2012) Saving by sharing – collective housing for sustainable lifestyles in the Swedish context, presented at the 3rd International Conference on Degrowth for Ecological Sustainability and Social Equity, Venice, 19–23 September, available at: http://www.slideshare.net/degrowthconf/saving-by-sharing-collective-housing-for-sustainable-lifestyles.

Van Schyndel Kasper, D (2008) Redefining community in the ecovillage. *Human Ecology Review*, 15, 1, 12–24.

Vestbro, D U and Horelli, L (2012) Design for gender equality: The history of co-housing ideas and realities, *Built Environment*, 38, 3, 315–335.

Walker, L (2005) *Ecovillage at Ithaca: Pioneering a Sustainable Culture*, Gabriola Island, Canada: New Society Publishers.

Walljasper, J (2010) *All That We Share: A Field Guide to the Commons*. New York: The New Press.

White, N (2002) *Sustainable Housing Schemes in the UK*, Nottinghamshire, England: Hockerton Housing Project.

Wilkinson, E (2013) Learning to love again: 'Broken families', citizenship and the state promotion of coupledom, *Geoforum*, 49, 206–213.

Wilkinson, E and Bell, D (2012) Ties that bind: On not seeing (or looking) beyond 'the family', *Families, Relationships and Societies*, 1, 3, 423–9.

Willem Duyvendak, J (2011) *The Politics of Home: Belonging and Nostalgia in Western Europe and the United States*. Basingstoke: Palgrave Macmillan.

Williams, J (2005) Designing neighbourhoods for social interaction: The case of cohousing, *Journal of Urban Design*, 10, 2, 195–227, p. 200.

Wimbush, P (2012) *The Birth of an Ecovillage: Adventures in an Alternative World*. England: FeedARead Publishing.

2 A politics of the common

Revisiting the late nineteenth-century Open Spaces movement through Rancière's aesthetic lens

Naomi Millner

Introduction

Today, notions of 'commons' take on new currency as forms of enclosure associated with corporate 'land-grabbing', and neo-liberal privatisation exerts increasing pressure on the possibilities for rural and collective livelihoods (Harvey 2011; Vasuvedan et al. 2008). Indeed, the longer histories and diverse geographies of (re)claiming commons can be thought of as a counter-current, or series of eruptive counter-currents, to the definition of private property in political economy and in law. However, thinking this way suggests commons as a binary opposite to enclosure – a polarity that I suggest is problematically reductive. Fresh articulations of the commons rely on the claims of earlier movements, but they rework claims, imagery and political agendas such that they perform a different spatial work, sometimes even producing new forms of regulation and containment. In scholarship, rather than being regarded as a universal given, I suggest that the commons must be understood through fine-grained studies which show up continuities, as well as discontinuities, within its mobilisation. In this chapter, I show how revisiting the late nineteenth-century Open Spaces movement through concepts of *aesthetics* can develop such nuance, whilst also opening new opportunities for politics.

The Open Spaces movement was an alliance of organisations that shared objections to the pace and extent of urban development which accompanied economic liberalisation. My study of the activities of one of these organisations – the Commons Preservation Society (CPS) – illustrates how the (re)claiming of commons is performed through constant efforts to frame and reframe open spaces in aesthetic terms. Such acts of claiming entail the production of new kinds of boundaries and spatial policing as much as the contestation of encroaching forms of enclosure. On the other hand, we can identify a 'politics of the common' in the perpetual excess of forms of common life to such boundaries and limits, and the ways in which this excess is rearticulated repeatedly into the public domain. Such rearticulations demand significant labour in order to avoid simply reiterating class divisions, nostalgic longings or dominant aesthetic sensibilities, or else they mobilise symbols and motifs which resonate *between* classes and orders. In terms of thinking about the present, I call for attention to this production of the 'common' within the study of the (re)claiming of commons. The common forms a horizon-line before and beyond specific spatial claims; it is a condition of belonging which exceeds group belonging and recalls instead the shared ontological conditions of human mortality. Such horizons cannot be captured – instead they haunt us. It is this haunting, especially as it invokes or characterises artistic forms of experimentation, which I shall suggest provides a vehicle for egalitarian politics in the present.

Critical to my reading of the CPS is the aesthetic perspective provided by political philosopher Jacques Rancière. A former student of Althusser, Rancière initially followed his master's interpretations of Marx to analyse the ideological dimensions of class oppression (Deranty 2010), but, after his participation in the occupations and demonstrations of 1968, he broke definitively with Althusser and orthodox Marxism, although not, for a while, with Marx himself (Hallward 2009). Rancière rejected Althusser's interpretations of Marx because they implied that oppressed classes were ideological victims who could not see through to the 'reality' of their situation. From this point onwards Rancière worked explicitly to unstitch the discourses of mastery and logics of 'hidden truths' that he identified in the structuralist project, as well as the widening gaps that he observed between theory and practice. In his early work this meant developing a commitment to the specificities of movements and the biographies of key protagonists within his approach to social history. As demonstrated in *The Nights of Labour* (Rancière 1989), this work drew attention to the 'backstage' work of small groups of working-class individuals in the making of political revolutions. Where orthodox Marxism names the proletariat and the party as forces of agency, we are invited to witness the importance of individuals' own self-education and acts of reclaiming.

This historical work informed Rancière's later theories of politics and aesthetics. Critically, Rancière (1999) has always insisted on a notion of 'politics' that is tied to the overturning of existing orders of seeing and saying the world. Such aesthetic revolutions can take place only when a 'part which has no part' – a group which is not only invisible but prohibited from political speech – makes itself of tangible account. Like the *sans-papiers* movement of undocumented migrants in France in the early 1990s, this means laying claim to social visibility and space in such a way that social roles and hierarchies are dramatically disrupted. It is here that the reason for focusing upon the aesthetic becomes clear. As modalities for experiencing and expressing the world (he focuses on notions of 'art') shift and change, so too must the way in which we locate the potentialities for disagreement and social reorganisation (Rancière 2006). Making this connection also helps scholarship to move beyond a mere diagnosis (and therefore reiteration) of class inequalities. Rancière's tactic is to prioritise sites and situations where claims to *equality* are being articulated in a way that makes them audible, where previously they were perceived only as noise. Such articulations can also be seen as acts of 'disagreement' (Rancière 1999) with the way that social parts are allotted – where disagreement is not a difference between white and black, but where two colours, both called white, are revealed to be qualitatively different things. Politics, for him, only really takes place when such disagreements take place in a public way, breaking with dominant forms of aesthetic 'policing', or recapture, that keep parts in their places.

In terms of commons, disagreement can therefore be regarded as an *aesthetic act of occupation*. The (re)claiming of commons is acted out in relation to felt worlds, through the appeal to artistic and literary sensibilities, because these are able to embody what has been forced to the margins of what can be seen and heard of the social world. Such claims cannot take place outside historical contexts, from which disagreement emerges, or without artistic tools and registers. Many members of the Open Spaces movement, for example, shared common literary and artistic influences in the Romantic painters and poets, especially in the ways these were being applied to social ideas by John Ruskin and William Morris. To 'materialise' the commons in these terms means revisiting the material spaces *of* commons with close attention to the historical contexts and sense conditions which made particular claims possible. Thus, the 'commons' claimed by the CPS is not the same configuration of rights and ideas as were defended either in the rural struggles of the seventeenth and eighteenth centuries, or those which animate today's Occupy movements. On the other hand, such movements did establish conditions for dissent which had a lasting effect on the ways that public space and the public realm are articulated in political terms. The mechanisms and practices associated with this production and its pre-history have also supported other, more 'radical' spatial claims, including those of the working-class movements of the early nineteenth century (Caffentzis 2010). Rancière's work suggests that enclosure and commons are not opposite or separable historical movements but *aesthetic dynamics*: opposed forces of policing and dissent which are continuously re-enacted through shifting social orders and artistic forms of creation. From this perspective the CPS embodied neither a force of pure politics nor one of aesthetic policing; instead it illustrates the contentious outworking of a new conception of public space.

The capacity of Open Spaces organisations to transform existing frameworks of seeing, saying and legislation was strongly associated with advocates' access to symbolic capital and spaces of decision making, which makes it problematic to think about in terms of equality. In the first section I review existing accounts of the CPS in these terms, highlighting aspects of material and aesthetic struggle which have been overlooked. In the two following sections I then draw on archival material to point beyond an analysis which limits politics to class dynamics. Although class remains an important analytical axis, I highlight shared religious, scientific and aesthetic sensibilities that also configured new social unities and affiliations. By situating the CPS in relation to the spatial and existential anxieties of the moment, I add new insights to existing accounts of this 'middle-class' improvement society. In particular, I draw attention to the new ways that the CPS constructed mechanisms for political disagreement within contexts where religion and traditional forms of authority were in decline and environmental transformations were fast paced. Finally, studying how claims and allies were assembled by the CPS will demonstrate the importance of identifying common *problems* within historical claims to space, rather than valorising common *solutions*.

The Commons Preservation Society: Brokers of the public good?

Protecting open spaces

The Commons Preservation Society is considered to be the oldest of the UK's national 'amenity' bodies – voluntary organisations set up for the purpose of preserving the art and architecture of the past. However, it was founded as an 'organised pressure group' (Cowell 2002) with the ambition to conserve and protect common lands threatened by the rapid privatisation of prior manorial properties during the late nineteenth century. The tensions surrounding the commons and its relation to liberty were felt especially in London, where liberal economics was quickly corresponding to reduced degrees of spatial freedom. In 1864 the tension around this redefinition of liberty and commons ran so high that when Earl Spencer proposed a Bill for the enclosure of Wimbledon Common, Frederick Doulton, the MP for Lambeth, was able to instigate a House of Commons select committee to re-examine the state of open spaces in London and to interrogate Spencer's Bill (Killingray 1994). In 1865 this select committee produced a report on London's commons calling for an immediate end to enclosure; the recognition of the rights of the 'public' over the commons; and the imposition of regulations for their better management. The CPS was founded in the same year with the purpose of giving effect to the select committee's recommendations.

George Shaw-Lefevre, the first Baron Eversley, was a key figure in the movement, and it was in his chambers in the Inner Temple on 19 July 1865 that the Society was inaugurated. Baron (or Lord) Eversley, the son of a Speaker of the House of Commons, was called to the bar in 1855. Entering Parliament in 1863, Eversley had become a prominent member of the radical group of Liberal MPs intent on reshaping the party for social reform in the context of the liberalisation of the corn trade (Williams 1965: 1). Leading up to the repeal of the Corn Laws in 1846, Parliament had been dominated by debates between well-represented landowners on the one hand, struggling to keep existing limitations on imports to ensure the premium value of their agricultural crops, and new industrialists on the other, keen to open up the markets and force down the cost of labour (Waller 1983). Eversley and his allies, including J.S. Mill, Lord Mount Temple, Professor Huxley, Sir Fowell Burrell, Charles Pollock and Andrew Johnston, developed a third-way argument within this debate, premised in ideas of a *public good*. Without rejecting the liberal ideas of individual ownership, they conceived of collective forms of ownership through which recreational land and cultural heritage could be protected from development, whilst also enshrining moral limits to the pursuit of individual gain. With £1400 contributed by Eversley's allies, the CPS quickly drew many eminent people into its projects, using 'commons' as a keyword to join their moral and spatial objectives.

Of course, concern over waning forms of common life was not new in the 1860s. The Chartist movement, which had been active between the 1830s and 1850s, had mobilised the term 'commons' in its occupation of public spaces such as London's Hyde Park, calling for rights to free assembly and rejecting the policing of public space (Hammond and Hammond 1930; Roberts 2001). Hedge-breaking, Leveller and Digger movements also actively claimed 'commons' when the manorial system of land management, dominant since Domesday, was being legally dismantled throughout the sixteenth and seventeenth centuries (Blomley 2007; Neeson 1993). This process continued into the eighteenth and early nineteenth centuries, although in the 1860s social movements were no longer responding primarily to the conversion of agrarian land into private property. Instead, the principal concern was the open land on the fringes of manorial estates that was being sold off for large urban building projects.

As a result of earlier popular resistance to these projects, in 1836 a General Enclosure Act included a public interest clause to prevent the enclosure of open fields within a ten-mile radius of London's centre. This politicised 'parks' in new ways: open recreational grounds, many of which were former common grazing lands of large estates, were framed as battlegrounds in the face of encroaching investors. Later, as well-known areas such as Hampstead Heath were threatened, the process was brought under the jurisdiction of the General Inclosure Act of 1845, which required the consent of one-third of 'commoners' (those using the land for turf-cutting, fishing, grazing, or regular access). The Act laid out a regulation process requiring parliamentary sanction (Baker 1940) and stipulated that, for commons near larger towns, allotments and field gardens must also be provided. However, between 1845 and 1869, 614,800 acres of common land were enclosed under approved orders, while only 4000 acres were allocated for public purposes (Eversley 1910; Williams 1965).

Historical enclosure within this context must be understood as a complex process, involving competing claims to conservation, protection and rights. For example, although the 1845 General Inclosure Act is often perceived as part of the process of establishing the well-ordered Victorian public park, the measures introduced were mainly deployed to slow down the new waves of building works making use of former agricultural land (Bailey 1987). There was a particular incentive for the upper classes to enclose during the 1860s and 1870s, with the import of large quantities of American wheat making the repeal of the Corn Laws finally tangible through a rapid decline in property values. However, the pressure applied by liberal spokespeople against such enclosures often conflicted with the views of 'commoners' on the use of space. 'People's' movements made the case *against* the 'People's Parks' being designed for them, arguing that the new conservation boards being set up to preserve them constituted another form of dispossession (Macmaster 1990; Conway 1991; Taylor 1995). The institution of park 'police' to ensure sanctioned uses of the commons was particularly opposed, being widely seen as an effort to regulate 'improper' behaviour in keeping with distinctively bourgeois norms (Rosenzweig 1984).

Materialising the commons consequently requires a 'thickened' conception of enclosure. Peter Linebaugh (2010: 11–12) writes that the English enclosure movement tends to be treated as a 'concrete universal', which, like the slave trade, witch burnings, Irish famine and the genocide of Native Americans, is rather glibly held to define the crime of modernity at large. Limited in time and place, but bound with possibility of recurrence, this enclosure is pitted against a commons which holds a 'promising but unspecified sense of an alternative'. Drawing on Linebaugh's words, McDonagh and Daniels (2012) suggest that this 'hold-all' concept consequently lacks sensitivity to the complex relations of space, power and ideas within a given historical moment. The works of historical critics, including rural poet John Clare, offer profound insight into other ecological and poetic possibilities of environmental inhabitation, but their reduction to a treatise for commons versus the matrix of neo-liberal privatisation gives rise to a 'thinned' and nostalgic conception of the commons. Meanwhile, experimental alternatives have always been multiple, fractured and various, and have often been accompanied by new kinds of spatial policing or unevenness. This is geographer Nick Blomley's (2007) emphasis when he revisits enclosure through the material construction of hedges. Whilst acknowledging the importance of symbolic acts, Blomley highlights the way that power relations are always materialised through *things* like fences, contracts and closed-circuit television cameras, which are also always subject to alternative uses (see also Brown 2001; Winner 1980). Thus, hedges were part of the spatial alienation of people from the land, but hedges could also be broken, used for firewood, hijacked by diverse eco-systems and materially renegotiated for continued access.

Whose public good?

From its inception, the CPS was characterised by both conservative and radical progressive tendencies. The grounds of its appeals were based in ideas of historical heritage – most notably, through the preservation of practices undertaken 'since time immemorial', whilst its models for the collectivisation of common life were inspired by experiments abroad and political critique. This tension is one reason why scholars have struggled to describe the CPS in terms of its class relations. Thus, Taylor (1995) and Allen (1997) portray the CPS as a conservative force, acting to preserve London's parks out of bourgeois desires to 'improve' the working class whilst ensuring their continued access to recreational spaces (see also Waller 1983). Likewise, Macmaster (1990) chronicles the resistance against a proposed park preservation scheme between the 1850s and 1880s on Mousehole Heath, a Norfolk site with particular symbolic significance, given its strategic location during Kett's fence-destroying rebellion of 1549. Macmaster indicts the Parliament-led commons-preservation scheme for its definition and suppression of 'rough' or 'unseemly' uses, as for its exclusion of long-standing working-class movements from their spatial control over the suburbs. Yet such scholarship downplays the radical reading of common and customary rights in the activities of the CPS and their role in stimulating popular dissent. It can also miss changes in popular sentiment across time. Macmaster finds that the movements 'from below' were ultimately successful in preventing the formalisation of a governing board of conservators through their mobilisation of effective symbolic capital – images and narratives which captured the affective weight of popular memories and galvanised widespread support. However, Mousehole Heath was one of the public parks whose regulated preservation the CPS eventually successfully advocated, with the support of local commoners.[1]

Other scholars underline the role of the CPS in facilitating new coalitions between classes, in many ways acting as the catalyst for the birth of new proto-types of early twentieth-century labour politics. Ben Cowell (2002), for exam-ple, underlines the continuation of a radical tradition of direct action in the CPS, describing the fence-break that it orchestrated in 1866 to defend Berkhampsted Common in Hertfordshire. In the event, 122 'navvies' from London were con-tracted to pull down iron fences enclosing approximately 430 acres in an act of occupation which Cowell associates with the development of an idiom of radi-cal intervention between social classes. Like Killingray (1994), writing on the dispute over Knole Park in Kent, and Eastwood (1996), describing the violent protests which followed and preceded the enclosure of Otmoor in Oxfordshire, Cowell views the legal and historical arguments employed by the CPS as a means of pitching on the side of the 'mob' in the defence of common practice. Such actions evidence the formation of social solidarities and the transcendence of usual class divisions.

The relationship between the CPS and the 'commoners' it claimed to side with is therefore far from clear cut, but I suggest that it can be further illuminated through an examination of its *eco-social* dimensions. The 'nature' produced by the CPS and its allies as they called for its protection is specifically a *socionature*: an entanglement of cultural ideas and processes with biophysical processes and sites, shaped by histories of social use as well as cultural interpolations of space and environment (Castree 1995; Swyngedouw 2006).[2] Revisiting socionatures through Rancière's terms of aesthetics demands a close engagement with the role of shared sensibilities in this production, and the ways in which artistic move-ments play into political responses. In the following section I treat two important themes which open up the politics of the common in this way: first, the notions of property and ownership which inflect social discourse about the environment within this context; and second, the notion of commons as 'commonwealth' which inflects property through a particular configuration of aesthetic sensibilities.

Common sensibilities

Commons and property

The tactics employed by the CPS made use of loop-holes within existing enclosure legislation to remove the grounds upon which wealthy landowners could claim rights to fence in and sell off their land. Dispute focused on the Statute of Merton, a legal instrument created as a form of compromise between Henry III and the barons of England in 1235 (Eversley 1875, 1910; Hunter 1884). This statute, which formed the basis for English common law, was initially structured to *allow* a lord of the manor to enclose common lands, provided that it could be evidenced that sufficient pasture remained for tenants. It was also the legal foundation upon which the aristocracy of the sixteenth and seventeenth centuries asserted their right to enclose (Neeson 1993). However, Lord Eversley's allies were well versed in legal intricacies, and the entire leverage of the CPS's actions was based on a sub-clause of the Statute, which stated that lords of manors could 'inclose' all or parts of the waste-lands of their manors *only provided* that it should appear on complaint of free tenants that a sufficiency of common remained to satisfy their rights. The solicitors and historians of the CPS mined local historical sources and local residents' memories in order to evidence a continuous use of the commons 'since time immemorial', and thus to object to enclosure plans on the basis that commoners' rights would *not* be satisfied. 'The idea', wrote Eversley (1910: 157), was 'to reverse the idea of absolute ownership of the Lords of Manors in the waste lands of their districts' and so to 'restore to the commons something of the attributes of the ancient Saxon folk land', thereby establishing the principle that 'they concern the interests of the people of the district, and the public generally, even more than the Lords of the Manors and their commoners'.

This rearticulation of the Statute in this way was enshrined in the General Inclosure Act of 1845 – an act that Lord Eversley considered to transform the Statute's very quality (Eversley 1910: 159). Eversley did not object to the enclosures of his day because they enclosed, but because they supported private gain over his passion: 'communal' interests. More is in play here than a simple reversal. The spatial imagination of the CPS is almost inseparable from the ideas of ownership and productivity that coalesced in this period, but it takes up the ideas of commons to expand a 'public good' from the private property relation. Specifically, Eversley's rhetoric marks the delineation of collectively held, or nationalised, property rights as a solution to the perceived excesses of liberal accumulation and development.

The strategic redeployment of the Statute of Merton this way was heavily influenced by ideas of property and rights which had been developing through the seventeenth and eighteenth centuries. Before the sixteenth century, property had been imagined as a bundle of overlapping and often non-exclusive rights and obligations (Thirsk 1981). Ideas of property and of ownership were linked with manorial boundaries and land deeds, although boundaries were constantly being renegotiated in relation to local customs and practice. This concept was transformed by John Locke and his disciples, who framed property as a 'bounded thing' (Blomley 2007: 5). Locke's *Second Treatise on Government* (1988), published in 1681, was a key document within this transformation, as it set property in relation to a new object of knowledge and field of intervention: the economy. Here 'commons' – a term previously denoting the 'wastes' of manorial estates – is reframed as the state of idle disuse which hypothetically preceded the instantiation of productive boundaries of ownership. Such boundaries were seen by Locke to transform labour by tying the fruits of labour to the land's owner, thereby introducing a motive for making the land work better. This *modus operandi*, which converts labour into 'value' (Locke 1988: 289, section 28), was set in contrast to systems of common management, which – Locke implies – encourage laziness.

In this move, ownership became tied to citizenship, and specifically to the notion of productive citizenship (Gidwani and Reddy 2011). Individual property, henceforth, was to be understood as a natural right that arises when individuals create value by mixing their labour with the land – a right which the market socialises when each individual gets back the value he or she created by exchanging it against an equivalent value created by another. However, individuals who fail to produce value have no claim to property. For example, the dispossession of indigenous populations in North America by 'productive' colonists was legitimised on the basis that they were allowing fertile land to go to waste and not cultivating from it the basis for a surplus (Parekh 1995).

For members of the CPS, these underlying principles of an adolescent liberalism were not exactly in question. The notion of communal, or public, ownership was their solution to the excesses of personal acquisition and environmental degradation which were becoming more apparent. E.P. Thompson (1975), whose social histories inspired Rancière's approach, makes a similar point when he argues for the 'order in this disorder' of earlier anti-enclosure movements which had seemed to reject claims to property. He finds that the traditional rights and legalistic rituals they resorted to, such as pledges of loyalty to the crown and acts of 'possessioning' land, were often themselves anchored in the vocabulary of property, rather than an affront to it. In such cases the effects of liberal industrialisation were being rejected, rather than its ontology. In the case of the Open Spaces movement, the concept of public ownership depends crucially on developing ideas of private ownership. The 'commons' advocated is not a rejection of private property, but an attempt to articulate communal interests into the law.

Protecting wild-lands

The adoption of Eversley's report on the grounds of the perversion of communal interests led to the passing of the Metropolitan Commons Act of 1866, which made enclosure of the London commons practically impossible. However, it also prompted lords of manors to rush ahead to enclose, beginning with Hampstead Heath. It was at this moment that, under Eversley's supervision, the CPS's first solicitor, Philip Lawrence, began to collect evidence of the continued existence of common rights on the Heath.[3] The battle for Hampstead Heath was concluded successfully in 1868, and was followed by protracted campaigns for commons, and later forests, including Wimbledon (1865–70), Wandsworth (1870–71), Plumstead (1866–71), Tooting Graveney (1868–71), Epping Forest (1865–71), Ashdown (1876–82) and Berkhamsted (1866–70). This sequence of interventions reflected a steady coalescence of the authority of the CPS to reshape the English countryside, especially on the fringes of cities. It also led to the development of a firm process to secure the intended outcomes. In each case the central CPS working group would bring a planned enclosure to local public attention, raising the issue in national papers and developing interest for a local CPS group. Meanwhile, the Society's key technicians – primarily historians and lawyers – would gather evidence to portray current uses of the land in question as a continuation of historical commoners' rights. A local resident with sufficient influence and funds – often a personal friend of Eversley's – would then be prevailed upon to bring a legal challenge against the lord of the manor, while public events were organised to rally the masses. CPS advocates usually ensured that the subsequent cases were heard in the Chancery rather than common law courts, since the discretionary authority of an equity court offered them a greater chance of success (Eversley 1910: 30).

What is significant for understanding the relevance of commons histories in this production of the public realm is the rejection, and in some cases reworking, of rational-actor models of economics. This was achieved through a quasi-mythical appeal to past natures. The appeal to 'time immemorial' allowed the productivity of citizens to be rethought beyond social-economic terms. Economic stability was not out of the equation, but 'good work' was being associated with a morality of 'higher' aspirations, said to be cultivated by exposure to beauty, a sense of connection with times past and sufficient leisure time. Public pleasure could consequently be deemed more worthy than private pleasure, but *also more likely to result in enduring yield*. Yet this publicly owned land was not wild, unmanaged land, and public pleasure did not entail unregulated social conduct. In keeping with its social function (to inspire and raise aspirations), it was to be characterised by strict ordering and moral codes. Lack of care for common environments was associated with other 'immoral' acts including sexual promiscuity, drinking and gambling. Conserving heaths, parks and forests threatened by building projects therefore also meant 'civilising' them. Civilisation, here, meant establishing limits to what could be touched by economic growth, while also reflecting a 'glorification' of nature – a reconnection between fallen humanity and its environment. Of course, this interpretation was neither uniform nor uncontested, as I shall aim to show.

However, in aesthetic terms the idea of commons was of course also closely tied to the 'wild nature' praised by the Romantic poets who inspired many CPS proponents. This led to an apparent sense of contradiction, since the socionature supported and advocated by the CPS was characterised by fencing, law, management and measurement, while ostensibly protecting reserves of wilderness from urban development. At Hayes Common, for example, which was in court in 1868, a successful 'defence' of the commons resulted in a manifesto for its regulation (CPS 1869). Tasks for the conservateurs, who could also levy rates on the commoners in the same manner as poor rates, included the drainage and improvement of the common, 'taking care to preserve its natural features', and to frame bye-laws and regulations 'for the prevention of nuisances and the preservation of order' (*ibid.*: 6).[4] Yet the idea of the 'natural' to be preserved through this process is key to understanding this seeming contradiction. The 'nature' of the commons produced was also human nature: generative, alive – perhaps the same nature that was being taxonomically recorded by Charles Darwin and his allies – yet with great potential for degeneration. Greed, sexual licentiousness and degraded environments were all read as signs of this ambiguous potential, notwithstanding the very different risks they posed to social stability. Nature needed society if it was to remain hospitable to society, while society needed nature to enshrine its 'higher' aspirations.

Key to the longevity and wide appeal of this framing was, thus, the way that commons as mythic imagination sutures together the 'social' and 'natural' worlds that are artificially separated in this equation. Here, protecting green spaces,

historical buildings and forests was part of the consolidation of a 'national inheritance' which extended backwards into the 'time immemorial' reiterated in the CPS
literature, and forwards toward social justice, through the transformative capacities of the soul stirred by nature. The myth of the national past, in what Gurney
(1997) calls the 'civic gospel' in landscape architecture, was oriented on the one
hand to transcendent ideals of communion with nature, and on the other to the concrete limits and project of national sovereignty. Through this move the nation itself
was artificially rendered 'natural' and permanent' – an ancient foundation with the
authority to demand allegiance. The English landscape of cleared spaces and trees
became the backdrop for paintings of reconstructed legendary scenes and stories,
reinforcing the aesthetic sense of 'nation' as weighty and historic. Rather than view
Open Spaces as the spatial imaginary *either* of bourgeois control *or* of prefigurative labour politics, we need, therefore, to understand the complexity of social tensions which set the aesthetic coordinates for new kinds of spatial claiming.

Common-wealth

Ruskin's legacy

The myth of open spaces promised to produce a healthy and sober working class,
upgrade the residential desirability of neighbourhoods and preserve the monuments of the past, at one and the same time that it frames all these transformative changes as a reconnection with the wilds of originary nature. But a myth in
many ways reflects a broader 'distribution of the sensible', as Rancière (2006: 40)
terms the way that the senses are policed, regulated and conditioned in a given
historical moment. The apparent contradictions in the aesthetics of Open Spaces
reflect shifts in scientific and religious sensibilities and different collective efforts
to renegotiate the stakes for common life.

The reading of nature consolidated by the CPS was strongly supported by
romantically inspired philosophical and literary figures of the nineteenth century,
including William Wordsworth, John Ruskin and William Morris. Engaged in
diverse social and artistic projects, many of these figures shared a reverence toward
nature in its power to move, uplift and inspire great acts, and a conviction that
social collectivity could be inspired through proper engagement with such powers
(Aitchison et al. 2014). The influence of Ruskin (1819–1900) was particularly felt
in this regard. Whilst Ruskin himself remained distant from the CPS, probably due
to his disagreements with J.S. Mill and his disciple Henry Fawcett (who became
a member of the CPS in 1869), Open Spaces pioneer Octavia Hill and her ally
Canon Rawnsley were close followers. Lord Mount Temple was good friends with
Ruskin, and the CPS's second major solicitor, Robert Hunter, was his acquaint-
ance. The disagreement between Ruskin and Mill is important: Ruskin opposed all
forms of utilitarianism and the *laissez-faire* economics inspired by Adam Smith
and Thomas Malthus, which were important to liberal thought of the day. Ruskin
argued that economics was a redundant science if it did not take note of the human
spirit and human aspirations, and, through his works on ecology, art, architecture

and social thought, espoused a vision of the incorporation of social affections bind-ing communities together into social institutions. Ruskin's commentaries on art provided a means to disagree with the partition of the environment according to the logic of profit and accumulation. They also channelled new ideas of natural process and dynamism from the sciences into the register of social change.

Ruskin's texts, and their rearticulation through authors and social thinkers including Morris, but also Leo Tolstoy, Oscar Wilde and Gandhi, were an impor-tant resource for the rearticulation of commons as a progressive social project. However, this influence points to an important tension in the aesthetic ambitions of the CPS. Ruskin's ideas negotiated connections between the laws of nature and laws of art, locating an emerging organic politics and co-operative social prin-ciple through an uneasy synthesis of influences, including Christianity, Natural Theology and Romanticism and modern science (Frost 2011a). This alliance of forces saw the dynamic conception of the material environment, popularised in recent schools of vitalism, competing with an essentially anthropocentric view of the environment derived from biblical exegesis. Man [*sic*] was both of, and sepa-rate from, the created world. Whilst explaining Ruskin's popularity amongst a generation situated between the claims of science and religion, this mixture would ultimately prove unsustainable (Frost 2011b). In the 1870s, faced by Darwin's more polemic account of interrelated species and temporal forces, Ruskin began to privilege an aesthetics of hierarchy and morality over his earlier conceptions of dynamic materiality. The CPS recapitulated many aspects of this vacillation, cele-brating the environment in vivid and prefiguratively ecological terms, yet always returning to frame their project within moral and reformist limits.

In this sense, Open Spaces can be seen as a ground of struggle between com-peting aesthetic sensibilities, which in turn were tied to competing models for knowing the living world. Commons were identified by the CPS with the agency and powers of change inherent to natural environments, but they were also prob-lematically associated with the recovery of a lost and more moral age. This met-onymic link is crystallised in a later work by Ruskin, his *Fors Clavigera* (1871), where he concretised his newer ideas of a cultural 'common-wealth' alongside his discussion of the merits of Old Communism.[5] This common-wealth was to be located in the remaining wild-lands of a country – whether managed forests and spaces or untended lands – together with the monuments of national herit-age and built remains of past civilisations. The conservation of such 'treasures' was associated with the production of a resource of aesthetic transcendence, sufficient to uplift souls and inspire them to higher forms of social life. Both everybody's property and nobody's property, the purposes of commons – as part of 'the public, or common, wealth' – was to be 'more and statelier in all its substance than private or singular wealth' (Ruskin 1871: 120). Yet, in the terms of Rancière's aesthetics, which refers to the way that ideas of art and artis-tic creation are entangled with articulations of common life, this 'higher' goal marks the violent imposition of one account of the social world over all possible others. The moral order of the CPS also designated parts and places without fundamentally rupturing the order of who can speak and what they can say.

On the other hand, employing Rancière's terms also makes clear how such forms of ordering, or aesthetic policing, are also closely connected with the negotiation of new terms of dissent. The connection of quasi-religious sensibilities with political economy in the language of the CPS signals an important transaction taking place between artistic, literary and scientific circles at this time. For example, in an extract of his seventh letter, 'Charitas', Ruskin's (Ruskin 1871: 120) 'old Communism' is used to drown the more individualistic tendencies of liberal capitalism, making it appear small minded and incomplete:

> And, finally and chiefly, it is an absolute law of old Communism that the fortunes of private persons should be small, and of little account in the State; but the common treasure of the whole nation should be of superb and precious things in redundant quantity, as pictures, statues, precious books; gold and silver vessels, preserved from ancient times [. . .] and vast spaces of land for culture, exercise, and garden, round the cities, full of flowers, which, being everybody's property, nobody could gather; and of birds which, being everybody's property, nobody could shoot.

Ruskin was not a socialist, although his work inspired many socialists, nor a Whig, though the emergent Labour movement of the early 1900s claimed him as one of its greatest inspirations. In many ways Ruskin's words here reflect the tensions between religion and science in this period, and cannot be easily associated with one social 'part' or another. In common with a number of CPS advocates, Ruskin identified with different movements and affiliations throughout his life, leading to contradictions within his own *oeuvre*. In 1858 he 'deconverted' from his Evangelical Christian faith as a result of his encounter with modern science and the biblical criticism of figures like Bishop Colenso, although he never abandoned his sense of 'spirituality' (Tate 2009: 118–119; Landow 1971). Ruskin's texts resonated among a generation confronting an absence of theological certainty, but seeking to continue its fidelity to an 'outside' to social life which would flesh out economic ideas of progress.

Without being a figure of what Rancière would term a radical articulation of 'disagreement' with a dominant aesthetic order, Ruskin's work inspired the social economy which characterised many later charitable, co-operative and non-governmental organisations (Maidment 1981). This attention underlines how a politics of aesthetics does not only entail the identification of a disruptive 'part' to a dominant order, as some interpretations of Rancière have suggested. Rancière's later works on film and literature make clearer that a politics of aesthetics also entails apprehending the *problems* being posed for common life in a given moment. Artistic media can both sediment given class orders and enable new kinds of collaboration. The critical point for understanding the (re)claiming of commons by the CPS movement is to identify how possibilities for common articulation were changing, and where the points of slippage from dominant to liminal orders can be identified.

Religion and recreation

This aesthetic reading of the commons is extremely important for understanding the motivations of the CPS and the wider appeal of the idea of Open Spaces. Elizabeth Baigent (2011: 31) notes that when such movements and their spatial effects are studied, explanatory factors of nostalgia, modernity and national identity are often taken into account but tend to be approached in definitively modern ways. Scholars do not take enough account of the framing influences on sense perception at the time, such as the 'sense of the divine rule and governance in life' which were 'almost instinctive' with a vast number of people (*ibid.*). Such theological sensibilities are particularly evident in the activities of the Kyrle Society, an institution set up Octavia Hill in sympathy with the Open Spaces mission in 1877. A close associate of Hunter, Hill established the Kyrle Society with the explicit motive of bringing 'the refining and cheering influences of natural and artistic beauty home to the people'.[6] Moved by the cramped living conditions and the new ailments of the labouring quarters she visited, Hill set up the Kyrle Society for the improvement of institutional spaces and city neighbourhoods according to the Open Spaces aesthetics and ideals. The Kyrle Society saw armies of predominantly female volunteers – against the more male-dominated CPS – decorating schools, parks and public spaces, constructing gardens and putting on theatrical activities. Hospital wards were adorned with images recalling the illustrations of Ruskin's friend and correspondent Kate Greenaway, such as the figure of a village child sitting in a wood, her lap filled with spring flowers. The activities of this Society evidence how the power of 'beauty', understood through romantic images and approaches to landscape, was increasingly understood to hold a power to move that was comparable to the transformative power of God. Such romantic imagery expands on the doctrine of incarnation, placing the child and the pastoral scene as a sign of nostalgia for a lost age of religious certainty, at the same time as a modern vision of a future age of spiritual fulfilment and freedom from material want (Offer 1981).

While easy to diagnose in the terms of moral improvement, what is interesting is that this aesthetic designation increasingly articulates social models in the terms of dynamism independent from the human imagination. In a moment where scriptures were being revisited with increasing poetic licence, there is an equivalency being made between the infilling of the spirit and the poetic encounter with environments. Hill's commitments to Open Spaces were premised in a conviction that the experience of beauty could 'save', but by this she meant, could transform and unify a vision for social life (Hill 1883). She hoped for social transformation, but saw its locus in an unregulated, personal experience of 'nature', whether as depicted in paintings or encountered in a park. Thus, aesthetics are central to this imagination of encounter, but there is a trusting confidence in the powers of this 'outside' to shape and guide. For Hill and her fellow Anglican and open space campaigner, Rev. H.R. Haweis, 'religion and recreation . . . meant to a certain extent the same thing – both meant to be born again' (Baigent 2011: 2). Nonetheless, the ideas of moral transcendence here are still transposed from the Christian heritage, along with implicit hierarchies of lower and higher states of nature. The mosaic of the poet George Herbert's words, 'All may have, if they dare try, a glorious life, or grave,' engraved on the walls of St John's Church, Waterloo Bridge Road by the Kyrle Society, make manifest these tensions.[7] Intended to 'give subject for thought to those who use the garden' and to associate the surrounding botanical ornamentation with an invitation to reflect on mortality, the motif blurs the boundary between the romanticist vision of poetic transformation in this life and the biblical promise of life after a worthy death.

Whilst the end of the nineteenth century saw the discrediting of much socioreligious work which had attempted to 'civilise' the Victorian poor, the Open Spaces movement was therefore precisely so influential because it appealed to members of a generation 'that was losing its Christianity but retaining a conditioned need for metaphysics' (Offer 1981: 335). Octavia Hill remarked in 1860 on the tendency of faith to 'cramp' life, but her architectural vision espoused renewed, undogmatic hope. The ideas of common-wealth, inspired by Ruskin and others, tying together the built environment with the green and living environment, also inspired the innovation of new institutions such as William Morris's Society for the Protection of Ancient Buildings,[8] and the National Trust. The latter, which still exists, was founded in 1895 by Octavia Hill, Canon Rawnsley and Robert Hunter for the purposes of acquiring property by gift or purchase for public ownership. As Lord Eversley had done in 1865, Hunter took advantage of the appeal and momentum of the Open Spaces movement to coalesce a new point of leverage: a commission with a mandate to compile a register of the earthworks, megalithic remains, ruins and buildings of historic or architectural interest in the country. This register would become a new basis for the development of the Open Spaces agenda, allowing action and acquisition to take place without the necessity of an enclosure dispute.

For the CPS, 'natural parks, over which every one may roam freely' were 'reservoirs of fresh air and health [which] bring home to the poorest something of the sense and beauty of nature' (Eversley 1910: 2). They were understood as national treasures and a public corollary to private property, which would inspire allegiance from citizens, admiration from abroad and self-improvement from the lower classes. The work of the Open Spaces movement was thus in one sense to spatialise the commons as new ritual centres of modern urban life. First green parks, but later playing fields, cemeteries and square-gardens, were designated the 'sanctuaries', 'sacred' and 'consecrated' places of cities' crowded populations (Hunter 1884: 5). Such places were described in letters between CPS members in the terms of 'pilgrimages' toward nature, who was their 'teacher' and 'deliverer'.[9] Thus, although the CPS did not play so great a role in later decades, the Open Spaces movement had lasting legacies. Not only did it succeed in materialising the 'commons' of the romantic imagination in the form of the English countryside, it also formed aspects of the legal architecture of public ownership and management, which persist to this day. Central to this architecture has been a reworking the vernacular of commons to include architectural, cultural and aesthetic assets as well as common grazing land. This sense of commons, along with implicit forms of nostalgia, has been critical to subsequent movements to (re)claim or 'occupy' commons. However, this sense of commons *on its own*, is not political; rather, it is fraught with multiple parts and parties, at different times disrupting and consolidating spatial rule.

Conclusions

The question of the aesthetics of acts of (re)claiming commons can also be rethought in terms of a politics of transforming relationships of authority. Authority did not vanish with the erosion of tradition, religion and traditional social hierarchies as mediators of social life. Instead, this moment of social reconfiguration saw the emergence of new institutions and practices bearing the old thread of connection to an 'outside' to social life into the new spatial problems raised by urbanisation. The mechanisms instituted through the CPS and the collective acts they staged – often in collaboration with others – can in this sense be understood as efforts to rearticulate the 'weightiness' of such an outside within social spheres increasingly saturated by the language and efficiencies of liberal political economies. Such acts established new forms of authority – new ways of claiming and political–legal practices through which to articulate – which drew not on traditional foundations but on principles of experience, evidence and testing. Whilst tending to reflect the moral cultures and interests of particular class groupings, such forms of authority were in theory open to anyone, and in theory premised in concepts of collective, rather than private, ownership.

Approaching such practices through concepts of aesthetics helps us to understand that the CPS embodied neither a force of pure politics nor aesthetic policing, but instead a contentious outworking of a new conception of public space. By materialising the commons we can identify the common political, economic and social problems under negotiation by the campaigns of the CPS, which were also problems of aesthetics. Responses, however, were fraught with contradictions supplied by resources drawn from scientific models, religious uncertainties and romantic myth. How might we conclude this study of contradiction and complexity? Particularly important for the project of connecting broader histories and geographies of (re)claiming commons is to understand what an aesthetic perspective adds to the study of dissent. For Rancière, 'politics' is an eruptive moment of spectacle, wherein a social party with 'no part' in the life of the community makes itself of some account. It is a historically specific instance where a stage of communication is constructed such that what was previously perceived only as noise becomes audible as clear claims to rights to presence. Approaching past political action through aesthetics means becoming attentive to such claims as they are assembled, and to attempts to police them into less disruptive forms. Rather than place the CPS on the side of enclosure or the side of commons, this calls for a nuanced understanding of the ways that forms of dissent were changing. To this end, we may highlight moments of 'commoning' which emerge from new sites of collaboration and hybridity, rather than 'looking for' the continuity of commons. Unlike 'commons', commoning is a verb, which implies the production of new modes of dissent and political subjects out of overlapping aesthetic registers and common sensibilities. Thinking acts of commoning helps us to identify emerging possibilities for collective political claims *together* with the assets which are to be preserved, rather than forcing a repetition of the commons/enclosure binary.

Commoning, in aesthetic terms, need not be a physical act of appropriation or a complete suspension of ordinary activities. Instead, it marks the theatrical moves in symbolic and artistic registers through which both of these forms of protest become possible and resonant beyond a particular social group. Thus, it is important to note that I am not suggesting that the CPS can be designated a subject of politics, in Rancière's terms. The CPS did not embody a 'party of the poor', and it was able to achieve success partly through the high concentration of cultural capital among key advocates. On the other hand, the spatial and symbolic claims articulated *through* the CPS did draw together unusual collectives and parties, as well as orchestrating moments in which its claims were rejected or reworked. This attention shows us how the 'public' sphere was being constructed as a political stage which reproduced the radical communication possibilities of literature and the arts. Whilst in practice utilised by some more than others, the emerging legal and spatial architecture of commons was technically open to anyone.

In this chapter I have followed in detail the ways that the claims of the CPS acquired validity and traction, laying out the backdrop of political-economic change, theological uncertainty and aesthetics which rendered its campaigns effective. I have identified key ways in which the activities of the CPS lastingly affected the legislation of the countryside, providing the basis for new institutions like the National Trust and creating new mechanisms for contention. On the other hand, I have also traced the new forms of spatial discipline which resulted from the developing ideas of public space and public parks, and the generation of increasing friction against other modalities for governing common life. Drawing these observations together with these concluding comments on politics, I would like to suggest that to take the struggles of the CPS seriously and *politically* is thus also to revisit their activities in relation not to the group's intentions, but to their effects. It is to revisit the archives of the past not so as to reduce the movement to one historical trajectory, but to notice specific moments in which dispute was made possible. From here it becomes possible to trace the material and aesthetic conditions of this possibility. To continue this politics into the present is, moreover, to continue to take part in scholarly and practical efforts to identify enclosure, where enclosure refers to the legislation of common life according to a metric of the profit of one – or the profit of some – over all others. The history of 'commons', as it is constantly being rewritten, can be seen as a history of attempts to revisit private property in these terms. Enacting a politics of the common means recognising that these attempts are multiple, various and both historically and geographically variegated. It is to grasp that in the struggle for the commons, more than one definition of common life is at stake.

Notes

1 Judgment regarding Mousehold Heath, 1883, printed, with manuscript note: to be reported by Commons Preservation Society, Surrey History Centre, Woking [hereafter SHC].
2 A parallel point is made in the field of Historical Ecology in Rackham (2000) regarding the interplay of natural and human in the creation of environments.
3 Bundle found wrapped in a copy of *The Times* for 18 December 1924, labelled 'HHE Parliament Hill' including printed 'Plea for the Extension of Hampstead Heath and the Preservation of Parliament Fields' by T.E. Gibb, Vestry Clerk of St Pancras, with plans and drawings, 1885, SHC; copy of the Hampstead Heath Act transferring Heath to the Metropolitan Board of Works [printed], 1871, SHC.
4 This early example of success in court produced a model which, it was hoped, would be extensively followed in respect of other commons. At Hayes the CPS was fortunate to have met with a 'willing' lord of the manor, but it was observed at this time that the principles being developed could in theory be applied even to cases where the lord refused his consent.
5 The 'common-wealth' is actually a fifteenth-century idea denoting public welfare, which had already been associated with the constitution of a democratic state. However, in this historical moment it was being newly used to refer to a wider community of justice – in 1884, for example, Lord Rosebery, a British politician, was one of the first to describe 'a Commonwealth of Nations'. See Queen Victoria's Jubilee Celebrations: Papers and correspondence concerning Victorian Open Spaces, 1896, SHC.
6 *Kyrle Society Founding Document* [printed] (1877: 2), in correspondence and papers of the Kyrle Society, including letters of Octavia Hill to Robert Hunter about leadership in the Open Spaces Committee (1884–1885), SHC.
7 *Kyrle Society Founding Document*, 4.
8 Whilst I lack the space to develop this theme here, William Morris's Society for the Protection of Ancient Buildings, formed in the 1870s, provides an interesting parallel to the CPS's later work to preserve heritage and properties. See W. Morris, *The Manifesto for the Society for the Protection of Ancient Buildings,* 1877, available online at: http://www.spab.org.uk/ (accessed 22 September 2013).
9 Correspondence of Robert Hunter on Borrowdale Road, and attack on Canon Rawnsley, 1912, SHC; Papers and letters to Robert Hunter regarding preservation of Petersham and Ham Commons, 1896, SHC; Open space in Red Cross Street, Southwark: letters and papers including copy of speech by Robert Hunter containing brief history of Red Cross Hall (1901–1906), SHC.

References

Aitchison, C., MacLeod, N. and Shaw, S. (2014) *Leisure and Tourism Landscapes: Social and Cultural Geographies.* London: Routledge.

Allen, R. (1997) The battle for the common: Politics and populism in mid-Victorian Kentish London. *Social History,* 22(1), 61–77.

Baigent, E. (2011) God's earth will be sacred: Religion, theology, and the Open Spaces movement in Victorian England. *Rural History,* 11(1), 31–58.

Bailey, P. (1987) *Leisure and Class in Victorian England: Rational Recreation and the Contest for Control, 1830–1885.* London: Routledge.

Baker, H. (1940) *Commons: What They Are and How They are Protected.* Watford: Commons, Open Spaces and Footpaths Preservation Society.

Blomley, N. (2007) Making private property: Enclosure, common right and the work of hedges. *Rural History,* 18(1), 1–21.

Brown, B. (2001) Thing theory. *Critical Inquiry*, 28,1–2.

Caffentzis, G. (2010) The future of 'the commons': Neo-liberalism's 'Plan B' or the original disaccumulation of capital? *New Formations, 69*, 23–41.

Castree, N. (1995) The nature of produced nature: Materiality and knowledge construction in Marxism. *Antipode*, 27(1), 12–48.

Conway, H. (1991) *People's Parks: The Design and Development of Victorian Parks in Britain*. Cambridge: Cambridge University Press.

Cowell, B. (2002) The Commons Preservation Society and the campaign for Berkhamsted common, 1866–70. *Rural History, 13*(2), 145–61.

CPS (Commons Preservation Society) (1869) *Report of Proceedings, 1868–9*. London: CPS.

Deranty, P. (2010) *Jacques Rancière: Key Concepts*. Durham: Acumen.

Eastwood, D. (1996) Communities, protest and police in early nineteenth-century Oxfordshire: The enclosure of Otmoor reconsidered. *British Agricultural History Society*, 44(1), 35–46.

Eversley, G.S.L.B. (1875) The Epping Forest Decision and its effect on the preservation of the commons. A letter to *The Times* and a speech delivered at the Town Hall, Shoreditch, by Sir William Harcourt. Reprinted for the use of the Commons Preservation Society, London, 1875.

Eversley, G.S.L.B. (1910) *Commons, Forests, and Footpaths*. London: Cassell.

Frost, M. (2011a) A vital truth: Ruskin, science and dynamic materiality. *Journal of Victorian Literature and Culture*, 39(2), 367–83.

Frost, M. (2011b) 'A strange coincidence [. . .] between trees and communities of men': The Law of Help and the formation of societies. *Nineteenth Century Prose*, 38(2), 85–108.

Gidwani, V. and Reddy, R. (2011) The afterlives of 'waste': Notes from India for a minor history of capitalist surplus. *Antipode*, 43(5), 1625–1658.

Gurney, P. (1997) The politics of public space in Manchester, 1896–1919. *Manchester Region History Review, 11*, 12–23.

Hallward, P. (2009) Ranciere's theatocracy and the limits of anarchic equality. In: G. Rockhill and P. Watts (Eds), *Jacques Rancière: History, Politics, Aesthetics*. Durham and London: Duke University Press, pp 140–157.

Hammond, J.L. and Hammond, B.B. (1930) *The Age of the Chartists 1832–1854. A Study of Discontent*. London: AM Kelly.

Harvey, D. (2011) The future of the commons. *Radical History Review*, 109, 101–197.

Hill, O. (1883) *Homes of the London Poor*. London: Macmillan.

Hunter, R. (1884) A Suggestion for the Better Preservation of Open Spaces, paper read at the Annual Congress of the National Association for the Promotion of Social Science, held at Birmingham, September 1884, reprinted for the Commons Preservation Society, London.

Killingray, D. (1994) Rights, 'riot' and ritual: The Knole Park Access Dispute, Sevenoaks, Kent, 1883–5. *Rural History*, 5, 63–79.

Landow, G. (1971) *The Aesthetic and Critical Theories of John Ruskin*. Princeton: Princeton University Press.

Linebaugh, P. (2010) Enclosures from the bottom up. *Radical History Review*, 108, 11–12.

Locke, J. (1988 [1681]) *Two Treatises of Government*. Cambridge: Cambridge University Press.

Macmaster, N. (1990) The battle for Mousehold Heath 1857–1884: 'Popular politics' and the Victorian public park. *Past & Present*, 127, 117–154.

Maidment, B. (1981) Ruskin, *Fors Clavigera* and Ruskinism, 1870–1900. In: R. Hewison (Ed.) *New Approaches to Ruskin: Thirteen Essays*. London: Routledge, 194–213.

McDonagh, B. and Daniels, S. (2012) Enclosure stories: Narratives from Northamptonshire. *Cultural Geographies*, 19(1), 107–21.

Neeson, J.M. (1993) *Commoners: Common Right, Enclosure, and Social Change in England, 1700–1820*. Cambridge: Cambridge University Press.

Offer, A. (1981) *Property and Politics 1870–1814: Landownership, Law, Ideology, and Urban Development in England*. Cambridge: Cambridge University Press.

Parekh, B. (1995) Liberalism and colonialism: A critique of Locke and Mill. In: J. Pieterse and B. Parekh (Eds), *Decolonization of Imagination: Culture, Knowledge and Power*. London: Zed, 81–98.

Rackham, O. (2000) *The History of the Countryside*. London: Phoenix.

Rancière, J. (1989) *The Nights of Labour: The Worker's Dream in Nineteenth-Century France*. Philadelphia: Temple University Press.

Rancière, J. (1999) *Disagreement: Politics and Philosophy*. Minneapolis: University of Minnesota Press.

Rancière, J. (2006) *The Politics of Aesthetics*: *The Distribution of the Sensible*. London: Continuum.

Roberts, J.M. (2001) Spatial governance and working class public spheres: The case of a Chartist demonstration at Hyde Park. *Journal of Historical Sociology*, 14(3), 308–336.

Rosenzweig, R. (1984) The parks and the people: Social history and urban parks. *Journal of Social History*, 18, 289–95.

Ruskin, J. (1871) *Fors Clavigera*. In: *The Works of John Ruskin, vol. VII*. Orpington: G. Allen.

Swyngedouw, E. (2006) Circulations and metabolisms: (Hybrid) natures and (cyborg) cities. *Science as Culture*, 15(2), 105–21.

Tate, A. (2009) 'Sweeter also than honey': John Ruskin and the Psalms. *The Yearbook of English Studies*, 391, 114–125.

Taylor, A. (1995) 'Commons stealers', 'land-grabbers' and 'jerry builders': Space, popular radicalism and the politics of public access in London, 1848–1880. *International Review of Social History*, 40, 383–408.

Thirsk, J. (1981) The common fields. *Past and Present*, 29, 3–25.

Thompson, E.P. (1975) *Whigs and Hunters*. London: Allen Lane.

Vasuvedan, A., McFarlane, C. and Jeffrey, A. (2008) Spaces of enclosure. *Geoforum*, 39(5), 1641–46.

Waller, P. (1983) *Town, City and Nation, 1850–1914*. Oxford: Oxford University Press.

Williams, W.H. (1965) *The Commons Open Spaces and Footpaths Preservation Society, 1865–1965*. Leicester and London: Commons, Open Spaces and Footpaths Preservation Society.

Winner, L. (1980) Do artifacts have politics? *Daedulus*, 109(1), 121–35.

3 A spirit of the common

Reimagining 'the common law' with Jean-Luc Nancy

Daniel Matthews

Introduction

The early-modern period witnessed a series of important shifts in English law away from medieval structures and customs and towards a putatively modern form of legality. With the rise of the modern nation-state came a concomitant desire for a centrally administered legal system that guaranteed the normative homogeneity of political space. It was the common law that emerged as the dominant legal force, supplanting or subsuming the plurality of jurisdictions that typified the medieval era. The *common* law was *common* to all within England and was a key instrument in consolidating the power of a newly centralised political authority. The early champions and apologists for this change to the nation's legal orderings – Coke, Plowden, Hale and, later, Blackstone – underpinned the common law's supremacy by reference to its ancient heritage, supposedly stretching to a time before not only the Norman Conquest but also the Roman invasion of Britain in the first century AD. This legal myth-making, as Peter Goodrich rightly points out, is a work of 'common law theology' (Goodrich, 1992: 205). By endowing the law with divine status, early-modern thinkers sought to guarantee the 'infallibility and unquestionable authority of an indigenous English law' (Goodrich, 1992: 205).

We might dismiss the narratives that claim that the heritage of the common law stretches to a 'time out of mind' (Coke, 1826: x–xii) as fanciful fables told by those seeking to acquire or sustain power. But, as with much forensic make-believe, these are operative fictions, instrumental in shaping the English legal and political imagination. Whilst lubricating the newly centralised machines of government, the common law came to occupy a unique position of power within the emerging modern state. As Coke was keen to remind James I, the king was thought to be beneath no one, only under God *and the law* (Coke, 1607: 65).[1] The common law's claim to a uniquely antique history guaranteed it as a locus of authority apart from – and arguably superior to – monarchy. Such a characterisation might inspire a 'classically liberal account of the strong common law as a safeguard against royal absolutism' (Cormack, 2007: 37). However, I argue, in what follows, that by attending to certain motifs and interests within the common law tradition we might open possibilities for reimaging the 'common law' that gesture towards an alternative political sensibility. Such thinking takes the 'common' to be the first question of law. In this account, law comes to be associated not with the liberal 'self' – the supposed bearer of our much-celebrated human rights – but with the on-going relations that constitute lives lived in common. This radical rereading of the 'common law' offers a corrective to the abstracted and reified language of rights by reawakening spirits of the common within the heritage of the common law. My thinking here hopes to counter the almost irresistible urge to associate law exclusively with state-sanctioned institutions and practices. The 'law' of the early-modern period meant much more than this; it gestured towards a 'law of community' and 'sociality' that preceded the exacting demands of positive or state law. By re-engaging this register of legality, I hope to articulate a sense of law which animates, rather than circumscribes, the social and communal bond.

The early-modern assertion of common law authority retained the medieval contention that 'law' referred to extant communal practices and customs undergirded by divine authority. These natural law justifications for the common law were prominent well into the nineteenth century before a backlash, led by Jeremy Bentham and others, paved the way for the positivism and formalism that dominated much legal thinking in the twentieth century (Bentham, 1948). The early common lawyers saw positive, or black-letter, law as a representation of some more fundamental lawful register. This important aspect to common law thinking, suggesting that law must be viewed as having an organic relation with the social and communal, is today sadly alien to many, even those working within the common law profession and academy. The early justifications for the consolidation of the common law spoke of 'law' not only in the register of statutes, cases and decisions. The common law imaginary of the early-modern period saw as integral to legal authority a sense of law as *ius*: the law as a bond that connects individuals to community, law conceived as part of the living architecture of social life.[2] The *ius commune* as the law of the land was, importantly, an *ius non scriptum*: an unwritten law, known and effective because it was 'inscribed invisibly on the heart . . . without the need for writing' (Goodrich, 2013: xxiii). In this register, the *common* of the common law assumes a different aspect. Not only referring to the administration of a common set of legal rules in a geographically determined area, the common law also evokes the 'co-' that is constitutive of the *commons*, *community* and *communalism*, a law that both expresses and answers to the bare fact that our lives are lived in common with others.

This notion of a 'deeper reality' (Postema, 1989: 3–4) which the early common lawyers saw themselves to be revealing in their judgments not only referred to long-practised rights and customs but also implicated every English subject through the category of 'allegiance'. Allegiance was a personal obligation that subjects owed to the monarch, guaranteed by both divine and feudal orderings. Referring to Coke's decision in *Calvin's Case* (1608),[3] Bradin Cormack suggests that allegiance for the early-modern lawyer was 'the ligament that connected minds or souls to one another' (2007: 246), a bond that united all subjects, through the sovereign. The notion of allegiance, for Coke in particular, was a radically personal matter, constitutive of the body politic but also operating on a personal and existential register.[4] As Cormack points out, allegiance was akin to a familial bond and inferred certain obligations by virtue of one's life within an existing network of relations (2007: 247). These obligations of allegiance and fealty preceded the positive law and provided an origi-nary social tie that, common to all subjects, was used as a prop for the common law's claim to authority. Allegiance operated as a fundamental ground for social relations and, by connecting each individual in a personal capacity to the authority of the sovereign, functioned as a key instrument in constructing and maintaining political power. Indebted to the rigid hierarchies of the feudal world and aspiring to the meta-physics of sovereignty heralded by modernity, allegiance pitches the social bond as a matter of stability and orderliness, positing an indubitable link between the individual and a transcendent source of authority.

To the contemporary mind, reeling from the Nietzschean deicide and awash with postmodern suspicions of claims to 'natural authority', this talk of obli-gation, allegiance and fealty is liable to raise hackles almost immediately. As Oliver Marchart (2007) has suggested, much contemporary thought is inspired by 'post-foundational' instincts and desires that directly challenge the categories that animated the early common law. For Marchart, post-foundationalism – common to a diverse set of thinkers from Derrida to Lefort, Badiou to Laclau – should not be confused with a spongy 'anything goes' post-modernism where foundations and foundational tropes such as ground, identity, totality, closure and so on are rejected *tout court*. Rather, contemporary radical philosophy – animated, in part, by a leftist inheritance of Heidegger – has sought to weaken grounds, founda-tions and essences so that they are seen as always already temporary and contin-gent (Marchart, 2007: 1–10). Such strategies, at first blush, are anathema to the early-modern thinking briefly outlined above. However, the post-foundational turn in much contemporary thought shares the early-modern insistence on the primacy of the social and – in particular iterations – like the early common lawyers, describes this prior sociality as a matter of 'law'.[5] In what follows I want to outline what is at stake in this thinking of a post-foundational account of the common and its relation to legality by reference to the work of French philosopher Jean-Luc Nancy.

Through Nancy's thinking, I hope to move towards a reimagination of a lawful register that precedes the positive law: a claim fundamental to the common law tradition but one that is often obscured or overlooked. Nancy's articulation of an ontology of being-in-common and his insistence on an existential dimension to obligation (that is, that *being* is always already a matter of *being-obligated*), offer resources for a radical reimagining of 'the common law'.

Rather than allude to allegiance and its concomitant thinking of fixity, pres-ence and sovereignty, Nancy's thinking of sharing, inoperativity and fragmenta-tion allows us to refashion law as that which is most proper to the common: law as an expression not of abstracted legalisms, norms and proscriptions but of the restless creativity of our being-with-others in the world. To be clear, my argument here does not advocate the reanimation of the feudal and natural law heritage of the common law. Such nostalgia for a supposedly lost communalism is anathema to Nancy's thought and, with its association with rigid social order and appeals to the infallibility of 'natural' or 'divine' justice, should find no place in our legal theorising today. Nor, it should be noted, do I seek to trace an undercurrent thinking of the common through the history of the common law, accurately plotting moments when such thinking resurfaces in the tradition. Rather, I hope to open a space where the 'spirit of the common', within the common law tradition, can be given its dues. This 'spirit' is rightly associated with the ghostly and spectral; it indicates an inchoate communalism buried (alive) within the common law tradition. As Jacques Derrida says of his engagement with a Marxian inheritance, such spectres should be 'maintained' and allowed to 'live' (Derrida, 2006: 1–3). To talk of spirits and haunting is, clearly, to be concerned with history, but a spectral history – unlike the claims of traditional historiography – 'is never given a date in the chain of presents . . . according to the instituted order of a calendar' (Derrida, 2006: 3). To speak to the spirit of the common law is to evoke a register beyond the assuredness of presence and linear temporal orderings, and it is, I argue in what follows, this ephemeral spirit, beyond a metaphysics of presence, which allows us to glimpse the promise and potential of 'the common law'.

Reimaging the common: community, fraternity and an ontology of the 'with'

Nancy explores the ways in which the common conception of 'community' – predicated on the deliberate or incidental exclusion of some in favour of others – obscures a more fundamental sense of our co-existence. We are beings, Nancy insists, who are always 'with' and this bare sense of our ontological 'withness' precedes the construction of particular communities or groups. Writing in *The Inoperative Community* Nancy (2008: 1–42) re-engages the category of 'community' as a vehicle through which an originary sociability can be recast in a world where such a sociability has been obscured. This post-foundational account theorises community as cohering not around some essence or common trait of belonging but around singularity and difference.

In its traditional rendering, community is aligned with a certain essence and identity: we are in such and such a community because of some identifiable trait of belonging. Nancy pursues a sustained critical engagement with such a conception. For Nancy, this 'traditional' notion of community has its roots in Rousseau's nostalgia for a lost sense of communitarianism. Rousseau casts society (state institutions and political regulation) as rupturing an originary community life of self-presence and stable identity:

> Rousseau . . . was perhaps the first thinker of community, or more exactly, the first to experience the question of society as an uneasiness directed towards the community, and as the consciousness of a (perhaps) irreparable rupture in this community.

> (Nancy, 2008: 9)

Here community is conceived in nostalgic terms whereby a harmonious and intimate communalism has been lost and needs, somehow, to be recaptured. Nancy suggests that this nostalgia lies at the heart of the perennial evocation of some golden bygone age where we were united in an organic communitarian cause. The task, for Nancy, is to interrupt these myths of essence, foundation and immanence and in so doing to think beyond communitarian models of community. In this very effort to think beyond the traditional parameters of community, Nancy re-appropriates the term for his own ends. Nancy distinguishes between 'communion' – a being-together conceived as a unity – and 'inoperative community' – a being-together orientated around difference that does not posit a unitary sense of togetherness (Nancy, 2008: 12). For Nancy, what is often referred to as 'community' reveals the totalising and unifying work of 'communion', and what is needed is an unearthing of the inoperative community that exposes a certain sense of our being together. Nancy works to reveal the constitutive sense of an inoperative community covered over by communion's unifying logic. Nancy contends that it is precisely the withdrawal of the essentialism of communion that is 'constitutive of "community" itself' (Nancy, 2008: 12).

It is perhaps here that the 'inoperativity' of the book's title comes into focus. Communion would be akin to an *operative* community where an essence is produced and put to work, where a myth of union is propagated and maintained. The 'inoperativity' (*désoeuvrement*) of community signals something radically different. This idiomatic term refers to idleness, unoccupancy, worklessness or uneventfulness, as well as un-working or rendering inoperative/non-functioning.[6] On the question of the *désoeuvrement* or 'unworking' of community Nancy suggests that,

> Community cannot arise from the domain of *work*. One does not produce it, one experiences or one is constituted by it as the experience of finitude. Community understood as a work or through works would presuppose that the common being as such be objectifiable and producible (in sites, persons, buildings, discourses, institutions, symbols: in short, in subjects) . . . Community necessarily takes place in what Blanchot has called 'unworking' [*désoeuvrement*] referring to that which, before or beyond the work, withdraws from the work, and which no longer having to do . . . with production . . . encounters interruption, fragmentation and suspension.
>
> (Nancy, 2008: 31)

The 'inoperativity' of the community is neither an absolute absence of work nor a project with a declared teleological orientation. For Nancy, the inoperative community is a 'gift to be renewed and communicated' and, whilst not produced through conscious 'work', is still a 'task', even a 'struggle' (2008: 35). The inoperativity of community is at once the condition of possibility for closed or essentialist communities (that is, 'communion'), but also the very force which interrupts and disrupts such communities.

Here we might ponder the question that Derrida pointedly raises in relation to Nancy's thinking. Whilst expressing a general sympathy for Nancy's 'inoperative communities' Derrida asks, 'why call them communities?' (Derrida, 2001: 25). Derrida's reticence is motivated by the danger of a totalitarian impulse lurking within the category of community. In *The Politics of Friendship*, Derrida ties community to a logic of 'fraternisation' whereby the *other* is transformed into the *brother*, where communal relations are governed by a paternalism and homogeneity, where difference is elided. In a community of brothers, or a community that evokes fraternity, the masculine authority of the brother (and by extension the son, husband, father) is privileged. Whilst discussing the relationship between fraternisation and community, Derrida comments 'there is still perhaps some brotherhood in Bataille, Blanchot and Nancy and I wonder, if it does not deserve a little loosening up, and if it should still be a guide for community' (2005: 48). Seemingly attesting to this charge, Nancy explicitly connects the inoperative community to fraternity, aligning 'fraternity' with an excessive and immeasurable notion of the sharing of our being-in-common, suggesting that 'fraternity is equality in the sharing of the incommensurable' (Nancy, 1993b: 72).

Derrida is right to be wary of Nancy's compulsion to retain the figure of the brother, even if purely symbolic. Why privilege the brother over 'the sister, the female cousin, the daughter, the wife or the stranger, or the figure of anyone or whoever' (Derrida, 2004a: 58)? Derrida rejects the use of 'fraternity' as a perpetuation of a Christian and psychoanalytic[7] privileging of the masculine figure of authority. And this concern colours Derrida's rejection of community. 'Fraternity' is rejected because the gendered, Christian and psychoanalytic heritage over-determines the concept's use but, Derrida argues, 'community' evokes a similarly fraternal spirit where each singularity is reduced to something common, a common *man* where difference is obscured. These concerns with Nancy's deployment of community have a direct bearing on our current efforts to reconfigure a notion of the 'common' in the common law. In Coke's formulation of allegiance, he directly evokes the familial – and patriarchal – bonds that exceed the positive law, suggesting that it is these 'natural' obligations and fealties to which the common law refers in order to guarantee its authority (Cormack, 2007: 248). Such evocations of the naturally arising, familial-cum-fraternal ties and obligations are the seeds of the very bad indeed. As Derrida counsels, at its root, Coke's fraternity names a totalitarian impulse that works to circumscribe and limit a sense of community at the very moment that it is presented as 'naturally' arising.[8] A reworked sense of the common in a post-foundational register must answer to the concerns that Derrida raises vis-à-vis Nancy. Being-in-common must be more than being-in-community, the common speaks to a connection that exceeds the jurisdictional logic that animates community's exclusionary impulses.

Acknowledging these concerns with community and fraternity, in more recent interventions on his ontology of being-in-common, Nancy has avoided the question of community, focusing instead of the barer category of the 'with'. Writing in 'The Confronted Community' (2003), Nancy reveals that since the publication of *The Inoperative Community* he has more acutely noted the 'dangers inspired by the usage of the word 'community'. Among these dangers, he notes the 'inevitable Christian reference . . . spiritual and brotherly community, communal community' (2003: 31). These difficulties prompt a privileging of the 'with' without the need to place this 'with' in reference to community or communion. This 'with', he writes, is

> almost indistinguishable from the 'co-' of community, it brings with it however a clearer indicator of the removal at the heart of proximity and intimacy. The 'with' is dry and neutral: neither communion nor atomisation, just the sharing and sharing out of space.
>
> (2003: 32)

This thinking of an ontology of the 'with', divorced from a logic of community, offers a radical rethinking of 'the common'. Developed in detail in *Being Singular Plural* (Nancy, 2000), Nancy's ontology is not concerned with the common as a question of inter-subjectivity, nor is it positing the common as something (indeed it is not a 'thing' at all) that is shared equally between all members in the sense of a proprietorial distribution. There is no subject that precedes the 'with' in Nancy's formulation: 'it is not the case that the "with" is an addition to some prior Being; instead, the "with" is at the heart of Being' (Nancy, 2000: 30). The notion of an inter-subjective relation (whether in phenomenological or psychoanalytic terms) presupposes a subject that – whether logically or historically – exists *before* relations with others. In Nancy's account the 'with' co-appears or 'compears' with each singularity (Nancy, 2008: 33), so that 'being' can be nothing other than 'being-with'. This 'with' is not something that each singularity *possesses*, rather it only appears *between* others; the 'with' is the sharing of being, rather than some thing that is shared out or distributed.

The common, in this formulation, is an ongoing series of relations that are constantly and simultaneously made and un-made. Maurice Blanchot captures this peculiar sense of inoperative connectivity by suggesting that if the relation between two singularities could be plotted it would take the form of a doubly dissymmetrical connection 'in which point A would be distant from point B by a distance other than point B's distance from point A' (2013: 66). In Nancy's terms, we can only *sense* such a relation.[9] As Blanchot's description testifies, this sense of a common relation exceeds the calculable, programmable and significative. In attempting to capture the common in the register of signification it withdraws, escaping the limits that we instinctively seek to impose upon it. Such a thinking of the common might a evoke a communism, 'without –*ism*' (Nancy, 2010: 150); a common without end or closure; the common as the on-going opening of being as being-in-relation. This is a common without essence or foundation – entirely different from the early-modern account of allegiance and natural fealties but, as we shall see in what follows, a radical thinking of the common that might allow us to recast the common law heritage anew.

Reimagining law: abandoned being and being-obligated

It is a failure of foundation and essence that animates Nancy's thinking of being as an expression of 'abandonment', and it is through abandonment that I want to turn to the questions of law and obligation. Nancy suggests that being must be thought of without reference to ground, essence or immanence. It is this failure of foundation that inspires Nancy's terminology – we are abandoned in the world and (particularly important for Nancy) abandoned in and amongst others. As just outlined, being is irredeemably 'with', but this 'with' can never be made into some foundation or ground. Nancy's use of 'abandonment' follows Heidegger's notion of *Geworfenheit*, which characterises *Dasein* as being 'thrown' into the world, unsuspecting of its lot and without request or pardon (Heidegger, 1962: 172–179).[10] Nancy's inheritance of Heidegger's *Geworfenheit* as 'abandonment' is significant, particularly in the way that it foregrounds legal themes in a discussion of the ontological. As with community, Nancy characterises being as something that cannot be captured but always withdraws. Both being and community are without essence or absolute ground: at the very moment that an essence of being/community is posited, being/community vanishes. In his essay 'Abandoned Being' (1993a) Nancy frames this movement of withdrawal in terms of abandonment, suggesting that the absence of an absolute ground for being means that being is simply abandoned: 'what is left is an irremediable scattering' (Nancy, 1993a: 39), the simultaneous withdrawal and fragmentation of being, as a unitary, stable, transcendental category. As Nancy, suggests, 'to be abandoned is to be left with nothing to keep hold of and no calculation' (1993a: 39). Abandonment, in this sense, should be understood as absolute withdrawal with no return, a permanent banishment from the metaphysical comforts of identity, fixity, permanence, unitariness and so on.

Nancy presents being as that which exceeds its own categorisation – indeed, Nancy contends that excess is the only modality of abandonment (1993a: 37) – but this excess is precisely what 'renews and revives' being (1993a: 42). Like the notion of inoperativity (*désoeuvrement*) that informs his thinking of community, abandonment marks the impossibility of foundation or ground, but is also seen to be that which animates the very possibility of the vitality, play and 'renewal' of life. Significantly, however, abandonment – unlike inoperativity – has a particular legal inference that is crucial for the present discussion of ontology, the common and law. Etymologically, abandonment is rooted in *bandon*, meaning 'jurisdiction, licence and control'. Abandonment implies being cast outside a particular jurisdiction, *ban*ished from the authority of law. However, Nancy insists that to be abandoned (which, remember, he claims is the very 'essence' of being) is not to be removed from law entirely. Rather, he suggests, 'one always abandons *to* a law' (1993a: 44), rather than being simply banished *from* a particular set of laws within a jurisdiction. Understanding this 'law' to which we are abandoned is crucial to the argument that I want to make. It is in Nancy's formulation of a 'law', to which we find ourselves subject merely by being, that we can begin to reimagine a radical sense of the 'common law'.

By allusion to the notion of banishment Nancy makes it clear that to be abandoned is not to be removed from law *tout court*. As Giorgio Agamben has discussed in relation the logic of the sovereign ban – and Agamben draws on Nancy in this respect – to be removed from the law (that is, to be banished) is to be subject to a relation of inclusive-exclusion (Agamben 1998: 58–62). To be banished or abandoned, then, is to be at once *outside* the law – or perhaps, more specifically, beyond the jurisdiction of specific laws – at the very moment that one is *subject* to the law in the most general sense. As Nancy puts it: 'abandonment does not constitute a subpoena to present oneself before this or that court of law . . . it is a compulsion to appear absolutely under the law . . . as such and in its totality' (1993a: 44). We should understand this condition of being under the totality of law to have an ontological inference. As Benjamin Pryor points out, law is a discourse that 'resists abandonment' (2004: 278) by constructing an image of itself as being immanent, closed and present. The expulsion from the law is akin to the failure of grounds and essence that is key to the notion of abandonment. So, to be abandoned (by law) is to be exposed to another law which is without immanence, presence or ground. This returns us to our concerns with community and Nancy's ontology of being-with outlined above. That to which we are abandoned is a law of plurality and difference, an ontological law that resists essence and identity. This is the law of the common, a law of 'being-with', a law necessary and unavoidable; as Nancy suggests, 'abandonment respects the law; it cannot do otherwise' (1993a: 44). To be is to be abandoned, to be without fixity, identity or presence; to be is to be cast outside oneself and exposed in an inoperative relation to others. We are, Nancy stresses, abandoned at birth and so are forced or compelled to be in a state of abandonment. This is 'the other of the law' (Nancy, 1993a: 44) that precedes juridical law and is revealed through a withdrawal of fixity or determinacy. In Nancy's formulation, abandonment from the law does not mean that we are abandoned to a particular place or space beyond the law, rather abandonment means being cast beyond such particularities. There is abandonment, and this is the law.

To suggest that there is a *law* of being-with or a *law* of the common that exposes selves to others is, perhaps, to use the term rather loosely. We might think of this law in the same way as the *law* of gravity or the second *law* of thermodynamics: that is, as a law that is (more or less) inescapable, or, at least, effective in given circumstances. But such *laws* might be thought to say little about obligation, injunction or compulsion. Let us examine a little further the way in which Nancy's ontological *law* of being-with is connected to these latter tropes; how does this ontological law obligate us in some way?

As we have already suggested, Nancy's 'abandonment' is indebted to Heidegger's *Geworfenheit*, the idea that being is a matter of being 'thrown' into the world. In Heidegger's account we are beings-in-the-world, coping with the exigencies of everyday life but without a clear sense of any foundation, origin, why or wherefore of our condition. This being-in-the-world, importantly, infers a 'must'. To be thrown into the world is to 'have-to-be' in the world at that particular moment in time and space. To be there (and *Dasein* is precisely this 'being-there') is to *have to* be there. There is a quiet 'ought', then, inscribed at the heart of being. In a recent essay that helpfully expands on and contextualises this notion of being-obligated, Francois Raffoul argues that key to understanding Nancy's take on this quiet 'ought' is the sense in which being must be conceived as responding to its own groundlessness. Abandonment infers that the grounds of our being resist final determination. In Raffoul's terms this groundlessness means that we cannot 'appropriate' a ground for being; that is, any supposed 'ground' will immediately withdraw if posited. As Raffoul argues, 'this inappropriability is for Nancy precisely – and paradoxically – what existence *has* to appropriate' (Raffoul, 2013: 73). The fact of abandonment, central to a post-foundational, Nietzschean inheritance to which Nancy is directly addressing, leaves us with a conception of being as being-abandoned. For Nancy, all that we can do – indeed, what we *must* do – is take this groundlessness of being as our 'ground' (the quotation marks are all-important here) to appropriate the inappropriability of our being in the world. This sense of being as being-abandoned 'desires itself and obliges itself' (Nancy, 1997a: 51), preceding from 'the nonavailability . . . of an essence (and/or power) of the self' (Nancy, 1993b: 28); or, as Raffoul puts it, 'obligation arises out of this groundlessness and becomes an obligation to this very abandonment' (2013: 75).[11]

Let us pause here in order to expand a little on the category of obligation. To be obligated means to be bound, with the Latin *ligare* providing the root for obligation, ligament, allegiance and so on. The fact of an obligation infers the possibility of enforcement; the ligaments of obligation tie two contracting parties, for instance, to their promise of performance. In its earliest formation in Roman law, obligation is first and foremost a question of debt (Holmes, 1881: 9–10). To be obligated is to be indebted to another. The significance of obligation and debt, for our purposes, turns on its role in shaping the laws of social and communal life. For Maurizio Lazzarato, who follows Nietzsche in contending that the obligations of creditor and debtor are at the root of all human social relations, 'the task of community or society has first of all been to engender a person capable of *promising*, someone able to *stand as guarantor for himself* in the creditor–debtor relationship, that is, capable of honouring his debt' (Lazzarato, 2012: 39–40). As Lazzarato carefully demonstrates in his important study *The Making of Indebted Man*, the mobilisation of this originary obligation of debtor–creditor has been fundamental in fashioning the social and political subjectivities and institutions of modernity. Nancy's insistence on the ontological dimension to obligation, however, speaks to a different register. Nancy's being-obligated is not a matter of interpersonal promises or questions of how social institutions have been constructed around this 'fundamental' relation. As we suggested, Nancy's being-obligated is a matter of being-obligated to/by abandonment; we are obligated to appropriate abandoned being as the very 'ground' for being itself. Such a turn recalibrates our thinking beyond the social and political institutions that manipulate subjectivities through the debtor–creditor relation to examine the nature of the bonds or ligaments that are prior to this. Through Nancy, we can see that the debtor–creditor couplet – supposedly 'grounding' the history of obligation in the West – is predicated on a prior relation, a bare facticity of being-in-common.

The obligation that Nancy evokes is an obligation in an ontologically radical sense. The ligaments that tie us to the common persist beneath, before or beyond any subsequent social and political constructions of indebtedness. If we follow Nancy in this register of the ontologically obligated, the nature of the social or communal bond changes form. We can no longer evoke the social bond as that which ties subjects to a particular political order or construction of community, in the sense that allegiance operated in the early-modern period. The obligation of being as an obligation to being-abandoned speaks of a bond that is beyond the rigidity of these structures. Being-obligated, in Nancy's terms, involves recognising the originary obligation generated by the failure of grounds. Derrida comes very close to Nancy in this respect when he suggests that:

> We are caught up, one and another, [bound-up, tied-up, obligated, perhaps] in a sort of heteronomic and dissymmetrical curving of social space – more precisely, a curving of the relation to the other: prior to all organised *socius*, all *politeia*, all determined 'government', *before* all 'law' . . . prior to all *determined* law, *qua* natural law or positive law, but not law *in general*. For the heteronomic and dissymmetrical curving of a law of originary sociability is also a law, perhaps the very essence of law.
>
> (2005: 231)

This essence of the law is an 'essenceless essence', an essence predicated on a paradoxical asymmetrical curving that exceeds the logics of exchange and precise plotting. To be obligated is to be abandoned to Derrida's curving of social space, we *have to be* caught up with one and another in this groundless relation. And this *has to be* of being is, as Derrida intimates, the very essence of the law.

If the nature of the social and communal bond changes form, in light of Nancy's thinking, so too does the sense in which we are originarily indebted. Following Nancy's insistence on thinking obligation through the register of the ontological, we might reconstruct debt in a similar fashion. We should think being-obligated and being-with not through the interpersonal creditor–debtor relation but through an indebtedness owed simply by virtue of our being in the world. Given Nancy's ontological commitments, such a debt would fall outside the realm of the calculable; it would signal a debt beyond measure, provoked by the immeasurability of abandoned being. To flesh this out a little, once again, Derrida helps. Evoking a sense of responsibility beyond the purely present, he suggests that we are responsible, 'before those who are not yet born or who are already dead, be they victims of wars, political or other kinds of violence, nationalist, racist, colonialist, sexist, or other kinds of exterminations' (Derrida, 2006: xviii). This is an exaggerated responsibility, a responsibility to and for that which we do not and cannot yet know, a responsibility for acts done without our knowledge, a responsibility beyond the living, beyond those present, beyond – surely – the human too. Where does one start with such responsibility? For Derrida the answer lies in inheritance. We are, from the off, burdened with a responsibility with and for the ghosts that we inherit simply by our being who we are: 'one is responsible before what comes before one but also before what is to come and therefore before oneself' (Derrida, 2004b: 5–6). It is in this sense that we are ontologically and structurally indebted. Not only does being hold within it a quiet 'ought', so too does being infer an inescapable indebtedness to the world into which we are thrown. This indebtedness prompts an originary responsibility or duty that Nancy describes, in his discussion of Heidegger's *ethos*, as an originary ethics of the 'non-proper' (1997a: 40). The groundless nature of being-singular-plural is not simply a given; it forces itself upon us and calls for a response, it takes the form of an absolute command or categorical imperative (Raffoul, 2013: 75–78). But this is an *unheimlich* or uncanny *ethos* (Nancy, 1997a: 41) because having-to-be-abandoned means that the grounds for our decisions become strangely serous under our feet.

This appeal to the realm of ethics, decision and judgement makes it clear that the sense of an ontological 'ought' that is most proper to the question of being is not simply a matter of being subjected to a law over which we have no control. As we suggested above, the condition of having-to-be-abandoned is a matter of taking the failure of the grounds of being as our very condition of being. This move that 'grounds' being on a groundlessness involves, in Nancy's terms, an act of creation, or more specifically a call for *the creation of the world* (Nancy, 2007). Creation, in Nancy's terms, is what we are left with if we embrace abandoned-being and acknowledge, as a point of departure for assessing ethico-political decisions and responsibilities, that any onto-theological ground for our decisions and judgements has withdrawn. Without a solid ground on which to construct our engagements with the world, we must create the world: 'in its profound truth, creation is thus nothing that pertains to a production or fashioning of the ground' (Nancy, 2007: 70). The creation of the world, then, involves a process of opening the world to what Derrida would call the *à venir*, to a future 'to come' that is without discernible *telos*.[12] The groundlessness of being means that predetermined ends should be exposed as mere mirages on a logocentric horizon. The obligation that Nancy identifies in being-abandoned calls for an opening of the world, an ongoing fashioning of human relations that must assume, as 'essential', the groundless being-in-common of our existence.

The promise of 'the common law'

During the period of its consolidation and rise to prominence, the common law sought to guarantee its authority by reference to the social and communal world. The common law was presented as an institution that spoke to both material practice as well as social and spiritual life. Writing in the early seventeenth century, Sir John Davies extolled the virtues of the common law, claiming that it was 'written only in the hearts of men . . . better than all the written laws in the world' (Davies, 1876: 253).[13] In this formulation, the law as *ius* – a social and communal set of practices known, and worthy of respect, simply by virtue of being in a community and living with others – outweighs the lofty proscriptions found in law books and statutes. As Goodrich suggests, underlining the importance of medieval orderings that retained currency in the early-modern period, the common law referred to a 'human geography which links the law to the inner life of the inhabitants of the various parts of the realm and to their local institutions' (1991: 258). The law was seen to be answering to and the product of people's creative efforts in the world. The medieval jurisdictions that the common law enfolded within its ambit were defined on the basis of the productivity of the land, marked by the inscriptions of 'the plough which dictates the boundaries of a village, a town or a city and its jurisdiction' (Goodrich 1991: 260). The ancient rights and customs that the early-modern lawyers sought to present themselves as protecting reached back to the laws of England before the Norman conquest. These were the laws enshrined in the Charter of the Forest of 1217 – the often overlooked counterpart to the much-vaunted Magna Carta – that protected the right to forage for food and have common access to the basic stuff of life. In the period which saw the common law come to prominence, we can see a centrally important appeal to the common, to the indubitable connection between the legal form and the creative life of people living and producing together.

How far we have come. Increasingly today, we are subject to the juridification of social life. Social and political obligations – solidarity, trust and the question of the common good – are obscured by the obsession with rights and the quasi-juridical management of social affairs. As Costas Douzinas argues, human rights have become the basis of almost all political claims and can be found on the lips of greedy consumers, left-wing radicals and forensic technocrats alike (2010: 1–21). This triumph of rights, however, has divorced the universal aspirations of freedom and dignity from our communal and social practices, presenting the legal subject as enisled from the social, a solitary and desiring individual rather than a communal and political animal. In biopolitical terms, law is often seen today as an instrument in the management of bare life, part of the administration of *zoe* through appeals to a just 'measure' or an abstracted notion of the 'normal' (Foucault, 2008). Such a conception of law circumscribes the common, cutting and dividing the onto-logical order to which Nancy's thinking speaks. This creeping juridification that sees law only as a regulatory force in a broader network of governmental control obscures the sense of law that animated the early efforts to consolidate the common law jurisdiction. The concern that law should express a constitutive relation with the common has come to be seen as an anachronistic fantasy of a bygone age, with its appeals to the quaint notions of allegiance and natural law. We are rightly suspicious of these categories that claimed an absolute ground for legal authority, but there is a certain sprit of the common in the English legal heritage that is wor-thy of our attention. Reworked through a radical displacement of foundations and grounds, a conception of 'the common law' as an ontological command to create the world anew offers tools for such a reanimation of this spirit of the common.

Where the biopolitical turn focuses on the measurable and seeks to administer the norms of bare life, Nancy's ontology engages a register of immeasurability, rearticulating the law as an injunction to create the world without appeal to the false certainties of foundational thinking. We – in our plural-singularity – must create the world by reopening the claims of *ius* within our common law herit-age. Such a reimagining of the common law involves rethinking the role of an originary *ethos* (and *ethos* signals both an ethics and also 'dwelling' and a sense of 'homeliness') within the common law. Reimagined within a post-foundational schema in the way we have suggested through Nancy, such an *ethos*, however, becomes strangely uncertain, an uncanny *ethos* without ground or ultimate *telos*. Such an endeavour seeks to return our thinking of law to the particularities – and peculiarities – of the common law heritage, and in so doing to reanimate a spirit of the common. This ephemeral spirit of the common cannot be reduced to an immanent or stable category. Rather, the common is precisely what resists such closure and, as I have suggested above, because of the constitutive relation that the common has with law, this reimagination of the common represents a chal-lenge to the staid legalism of contemporary rights discourse.

Through Nancy, we can open these two categories of 'common' and 'law' to new vistas. At the level of the ontological, the common becomes the law. The common is a *must*: we are obligated in *medias res* by the groundless relation of being-in-common. This *must* of the common is not only the first question of being but also the first question of law. This originary legality, co-appearing with being-in-common, poses itself as an injunction or command: a command to create the world, to embrace the fact of our abandoned-being and see our engagements in the world, whether in the spheres of politics, friendship, community or law as *creating the world*, an on-going production, lacking grounds or ends, opening possibilities 'without model and without warranty' (Nancy, 1997a: 158) for a renewed sense of law and justice.[14] Recreated, in light of Nancy's thinking, we might venture that this is the promise of the tantalising, but much maligned, syntagm 'the common law'.

Acknowledgments

Many thanks to Scott Veitch, Stewart Motha and Tara Mulqueen, who provided constructive feedback on earlier drafts of this piece.

Notes

1 Quoted in Margaret Davies (2002: 45). My emphasis.
2 *Ius* refers to a kind of informal, customary normative architecture that is commonly known and understood by a particular community or collective. This should be distinguished from law as *lex*, which is generally considered to have its root in *legere*, 'to read', and so is associated with written and substantive law. For a thorough discussion of *ius*, particularly in early Roman law, see Peter Stein (1966).
3 *Calvin's Case* (1608) 7 Co Rep 1a, 77 E. R. 377.
4 Coke mobilised a personalised understanding of allegiance to justify the reach of the common law to Scottish subjects after the succession of James VI of Scotland to the English throne in 1603, in *Clavin's Case* (1608). For a discussion of this case and the significance of Coke's construction of allegiance see Price (1997).
5 The use of 'law' to refer to an ambivalent sense of juridical law and the condition of possibility for juridical that overflows or exceeds this categorisation is common to Derrida (1992: 230–300), Blanchot (1988: 24–26) and Foucault (1990: 7–58).
6 For a helpful discussion of the English translation of *désoeuvrement* see Pierre Joris's 'Translator's Preface' to Blanchot's *Unavowable Community* (1988: xxi–xxv).
7 Nancy refers to Freud's patricidal myth of the origin of the moral law in *Totem and Taboo* during the discussion of fraternity; the 'sharing of the (dismembered) body' evokes a Christian as well as psychoanalytic heritage. Nancy contends that fraternity is 'the relation of those whose *Parent*, or common substance, *has disappeared*, delivering them to their freedom and equality. Such are, in Freud, the sons of the inhuman Father of the horde: becoming brothers in the *sharing* of his *dismembered* body' (Nancy, 1993b: 72).
8 It is arguably this logic of fraternity within the common law tradition – with its necessary logic of exclusion – that underpinned the common law's role within the colonial context. In fact, *Calvin's Case* (1608), in which Coke elaborates on his 'fraternal' understanding of allegiance, can be read as one of the first imperial cases in the common law tradition (Price, 1997). The exclusionary thinking at the heart of the decision is what a reimagined 'common law' must get beyond.
9 Nancy distinguishes between signification and sense. Sense is that which exceeds the closed system of signification. See Nancy (1997b: 12–34).
10 As Nancy suggests, Heidegger's 'being-there' (*Dasein*) implies 'a "duty of being-there", which is to say immediately the abandonment of existence to an obligation' (1993b: 26).
11 It is worth noting here that the connection between abandonment and obligation conditions Nancy's account of freedom. In *The Experience of Freedom* Nancy privileges the figure of the pirate as an exemplar of this groundless, abandoned, freedom. The pirate is groundless: neither answering to state, nor to some other higher authority, the pirate lives their life on the *chora* of a lawless sea, without jurisdiction. The pirate, nonetheless, is 'given over to the peril of [his] own lack of foundation' (Nancy, 1993b: 20), the pirate, therefore, takes his lack of grounds as their very ground – as Nancy suggests, piracy always implies some 'foundation' (Nancy, 1993b 85). It is this piratical freedom that Nancy champions, and it is this acceptance of groundlessness that informs the 'obligation' that co-appears with being.
12 Derrida makes the important distinction between 'the future', referring to a predictable future predicated on the present and 'the to come' (*à venir*), referring to a temporality beyond the present. The *à venir* is connected to an unforeseeable coming, tied to the logics of the event.

13 Quoted in Goodrich (1991: 259).
14 Nancy's evocation of possibilities 'without model or warranty' is picked up by Illan rua Wall (2012: 133–146). In his reimagining of human rights and constituent power, Wall draws on Douzinas's 'right-ing' (briefly developed in *The End of Human Rights*) and Nancy's understanding of 'creation' and 'invention' in theorising an 'open constituent power' devoid of teleological commitments to constituted power.

References

Agamben, Giorgio (1998) *Homo Sacer: Sovereign Power and Bare Life,* Stanford: Stanford University Press.

Bentham, Jeremy (1948) *A Fragment on Government with an Introduction to the Principles of Morals and Legislation,* Oxford: Blackwell.

Blanchot, Maurice (1988) *The Unavowable Community,* translated by Pierre Jorris, New York: Station Hill Press.

Blanchot, Maurice (2013) *The Infinite Conversation,* translated by Susan Hanson, Minneapolis: University of Minnesota Press.

Coke, Edward (1607) 'Prohibitions del Roy, 12 Co.', *Rep* 63: 65–77.

Coke, Edward (1826) *The Reports of Sir Edward Coke, Knt: In Thirteen Parts,* Volume II, J. Butterworth and Son.

Cormack, Bradin (2007) *A Power to do Justice: Jurisdiction, English Literature and the Rise of the Common Law 1509–1625,* Chicago: The University of Chicago Press.

Davies, Margaret (2002) *Asking the Law Question: The Dissolution of Legal Theory,* 2nd edn, Sydney: Lawbook.

Davies, Sir John (1876) *A Discourse of Law and Lawyers,* Dublin: Frankton.

Derrida, Jacques (1992) *Acts of Literature,* edited by Derek Attridge, London: Routledge.

Derrida, Jacques (2001) *A Taste for the Secret,* translated by Giacomo Donis, Cambridge: Polity.

Derrida, Jacques (2004a) *Rogues: Two Essays on Reason,* translated by Pascale-Anne Brault, Stanford: Stanford University Press.

Derrida, Jacques (2004b) *For What Tomorrow . . . : A Dialogue,* Stanford: Stanford University Press.

Derrida, Jacques (2005) *Politics of Friendship,* London: Verso.

Derrida, Jacques (2006) *Spectres of Marx: The State of Debt and the Work of Mourning,* translated by Peggy Kamuf, Oxford: Routledge.

Douzinas, Costas (2010) *The End of Human Rights: Critical Legal Thought at the Turn of the Century,* Oxford: Hart Publishing.

Foucault, Michel (1990) *Maurice Blanchot: The Thought from Outside,* London: Zone Books.

Foucault, Michel (2008) *The Birth of Biopolitics: Lectures at the College De France, 1978–1979,* London: Palgrave MacMillan.

Goodrich, Peter (1991) 'Eating Law: Commons, Common Land, Common Law' *The Journal of Legal History,* Volume 12, Issue 3, pp. 246–267.

Goodrich, Peter (1992) 'Critical Legal Studies in England: Prospective Histories', *Oxford Journal of Legal Studies,* Volume 12, issue 2, pp. 195–236.

Goodrich, Peter (2013) *Legal Emblems and the Art of Law: Obiter Depicta as the Vision of Governance,* Cambridge: Cambridge University Press.

Heidegger, Martin (1962) *Being and Time,* translated by John Macquarrie and Edward Robinson, London: Blackwell.

Holmes, Oliver Wendell (1881) *The Common Law,* Boston, MA: Little, Brown.

Lazzarato, Maurizio (2012) *The Making of Indebted Man,* Los Angeles: Semiotext(e).

Marchart, Oliver (2007) *Post-foundational Political Thought: Political Difference in Nancy, Lefort, Badiou and Laclau,* Edinburgh: Edinburgh University Press.

Nancy, Jean-Luc (1993a) *The Birth to Presence,* Stanford: Stanford University Press.

Nancy, Jean-Luc (1993b) *The Experience of Freedom,* translated by Bridget McDonald, Stanford: Stanford University Press.

Nancy, Jean-Luc (1997a) *Re-treating the Political,* edited by Simon Sparks, London: Routledge.

Nancy, Jean-Luc (1997b) *The Sense of the World,* Minneapolis: University of Minnesota Press.

Nancy, Jean-Luc (2000) *Being Singular Plural,* translated by Robert D. Richardson and Anne E. O'Byrne, Stanford: Stanford University Press.

Nancy, Jean-Luc (2003) 'The Confronted Community', translated by Amanda Macdonald, *Postcolonial Studies,* Volume 6, Number 1, pp. 23–36.

Nancy, Jean-Luc (2007) *The Creation of the World or Globalization,* translated by Francois Raffoul and David Pettigrew, New York: State University of New York Press.

Nancy, Jean-Luc (2008) *The Inoperative Community,* translated by Peter Connor, Lisa Garbus, Michael Holland, and Simona Sawhney, Minneapolis: University of Minnesota Press.

Nancy, Jean-Luc (2010) 'Communism, the Word' in *The Idea of Communism,* edited by Costas Douzinas and Slavoj Zizek, London: Verso.

Postema, Gerald (1989) *Bentham and the Common Law Tradition,* Oxford: Clarendon Press.

Price, Polly (1997) 'Natural Law and Birthright Citizenship' *Yale Journal of Law & the Humanities,* Volume 9, Issue 1, pp. 73–146.

Pryor, Benjamin (2004) 'Law in Abandon: Jean-Luc Nancy and the Critical Study of Law' *Law and Critique,* Volume 15, Issue 3, pp. 259–285.

Raffoul, Francois (2013) 'Abandonment and the Categorical Imperative of Being' in *Jean Luc Nancy: Justice, Legality and World,* edited by Benjamin Hutchens, London: Bloomsbury.

Stein, Peter (1966) *Regulae Iuris: From Juristic Rules to Legal Maxims,* Edinburgh: Edinburgh University Press.

Wall, Illan rua (2012) *Human Rights and Constituent Power: Without Model or Warranty,* Abingdon: Routledge.

Part II
Commoning

4 The more-than-human commons

From commons to commoning

Patrick Bresnihan

The question of the commons

Arguably, the most influential perspective on the 'commons' has been the institutional approach to managing 'common-pool resources' (CPR), a field of research that garnered Elinor Ostrom, one of its best-known proponents, a Nobel prize in 2008 (Acheson and McCay 1990; Hanna et al. 1996; Ostrom and Schlager 1992; Ostrom 2000). In the introduction to her seminal text, *Governing the Commons*, Ostrom makes it clear where the motivation for researching different rules, norms and institutions for collective action lay: in the power of certain dominant metaphors for explaining the causes of resource degradation and the limited choice of management strategies that emerged from such explanations.[1] Ostrom's great contribution is in challenging such metaphors. Rather than being helpless in the face of resource problems, Ostrom describes how individuals in very different parts of the world have come together to collectively devise and implement rules that have proven successful in terms of sustainably managing CPR, such as forests, fisheries and land. Based on these examples, certain principles and conditions favourable for producing institutions of community-managed resources have been identified and incorporated into management policies; throughout the 1990s and 2000s, models of community-based resource management have proliferated as an important policy instrument for the governance of natural resources (Agrawal 2003; Leach 2008; Li 2006).

The CPR approach rightly seeks an alternative to the powerful and reductive narratives derived from liberal and neoliberal economic theory. The problem is that in the rush to devise and implement rules and institutions for community-based resource management some of the defining features of these narratives can be reproduced. The need to remedy the 'distortions' of unregulated access to a resource, to transform it into a 'common property regime', is predicated on the same liberal identification of rational, economic subjects exploiting limited stocks of bio-physical resources. In other words, the CPR perspective begins with the same 'naturalized' assumption as do the 'tragedyists': without proper rules and norms, individuals will degrade and ultimately destroy common resources (Goldman 2004). Central to this formulation is the continued need for some form of property right as a means of shaping individual behaviour

(Dietz et al. 2003). This perspective perpetuates the 'methodological individualism, self-interested rationality, rule guiding behavior and maximizing strategies' that one would associate more with neoclassical economic perspectives on resource management (McCann 2004: 7). CPR approaches thus tend to normalize and naturalize particular bio-economic subjectivities and 'natures', ignoring the historical emergence of capitalist relations of (re)production in particular contexts, as well as the complexity of non-capitalist social relations, subjectivities and practices that exist alongside these (Mansfield 2004).

A second perspective on the commons that has become popular within and outside the academy shifts attention away from the so-called 'natural' commons, focussing instead on the emergent possibilities of the 'social' or 'immaterial' commons. These include the knowledge and cultural commons (Hyde 2011), the digital commons and peer-to-peer production (Bauwens 2005) and the biopolitical commons (Hardt and Negri 2009). While the political perspectives that inform these analyses differ, they all assume an analytic distinction between the 'immaterial' commons and the 'material' commons. In his article 'Two Faces of the Apocalypse', for example, Michael Hardt describes the difference between anti-capitalist activists and climate change activists at the United Nations Climate Change Conference (COP 15) in Copenhagen (Hardt 2010). While the former insist that 'another world is possible', the latter adopt the slogan: 'There is no Planet B'. Hardt traces these different political positions to their contrasting notions of the commons. On the one hand, anti-capitalists consider the commons as a social/economic commons, representing the product of human labour and creativity, including ideas, knowledge and social relationships. On the other, environmental activists speak for the ecological commons, identified as the earth and its ecosystems, including the atmosphere, rivers, forests and forms of life which interact with them. Hardt argues that the former does not operate under the logic of scarcity, while the latter does.

While the first perspective on the commons emphasizes the 'natural' resources on which we all rely, the second emphasizes the 'social' resources that have become increasingly central to contemporary forms of capitalist accumulation. In the first case, 'nature' (commons) is a stock of bio-physical resources which, as Hardt identifies, is subject to the logic of scarcity, bringing us into the domain of liberal political economy and the institutions of formal and informal property rights. In the second, 'nature' is no longer represented as a material background limiting human activity but becomes something malleable and infinitely reproducible, subject to recombinant technologies and human creativity. This is the domain of neoliberal political economy and the fantasies of contemporary capitalist (re)production (Cooper 2008). The problem with this distinction is that we end up with one form of the commons that appears to be asocial (excluding the socially productive and reproductive labour of humans involved in caring for the 'natural' resources they rely on), and another that appears to be anatural (excluding the material limits and properties of more-than-human bodies involved in the (re)production of the 'social' commons). While the distinction between the material/natural commons and the immaterial/social commons can be analytically helpful it tends to be over-stated, obscuring the continuity and inseparability of the material and the immaterial, the natural and the social.

A third perspective on the commons does not admit such a distinction and thus takes us in a different direction. From feminist scholars (Federici 2001; Mies and Bennholdt-Thomsen 1999, 2001; Shiva 2010; Starhawk 1982), geographers (Blomley 2008; St Martin 2009) and historians (Barrell 2010; Linebaugh 2008, 2011; Neeson 1996; Thompson 1993), we learn that the commons was never a 'resource'. The commons is not land or knowledge. It is the way these, and more, are combined, used and cared for by and through a collective that is not only human but also non-human. That the commons can continue to be identified as a 'resource' and not as a complex of relations between humans and non-humans attests to the long history of invisibility associated with 'nonrepresentational, affective interactions with other-than-humans' (De la Cadena 2010: 346). The 'invisibility' of peasant and indigenous cultures and forms of life has been well documented by historians and anthropologists (Brody 2002; Rose 2006; De la Cadena 2010; Escobar 1995; Linebaugh 2008; Thompson 1993); colonialism begins with the erasure of any existing claims to territory or history on the part of those who are being colonized. The concept of *terra nullius* refers to the identification of 'waste' land, or land that has not been inscribed with human culture and production. This term was not just used in the conquest of territories in the 'New Worlds' but also in the enclosure of common lands, moors and heaths that took place in Britain during the eighteenth century (Goldstein 2013). Silvia Federici, for example, argues that enclosure relies on the epistemological separation of the social and the natural spheres, the productive and the reproductive. She reads this separation-through-enclosure as something far more fundamental than simply the privatization of land. The relegation of 'women's work' (childbirth, child rearing, cleaning, cooking, caring) to the domestic sphere outside of the 'productive' economic sphere represents the 'naturalizing' of this kind of labour: '[a]ll the labour that goes into the production of life, including the labour of giving birth to a child, is not seen as the conscious interaction of a human being with nature, that is a truly human activity, but rather as an activity of nature, which produces plants and animals unconsciously and has no control over this process' (Mies 1998: 45). While reproduction is most often associated with human reproduction and the management of the 'household', from childbirth, to childcare and healthcare, cleaning and cooking, reproduction also extends beyond the confines of the house narrowly construed as four walls. Federici herself describes how her time in Nigeria observing and documenting the labour and activity of women in mostly subsistence economies led her to extend the notion of reproduction (Federici 2012): the household, or *oikos*, was not just a home or family but a wider sphere of communal reproduction that involved direct relations with the land, water, plants and animals, for example.[2]

The conclusions that are drawn from these insights are that capitalist enclosure and biopolitical control necessarily involve the de-valorizing and 'invisibilizing' of those myriad, situated relations and practices of (re)production that exist between people and the manifold resources they rely on (De Angelis 2007; Federici 2001; Shiva 2010). What is significant is that this understanding of the commons focuses on the particular relations and practices that characterize the commons as a different mode of (re)production. As Peter Linebaugh explains, '[t]o speak of the commons as if it were a natural resource is misleading at best and dangerous at worst, the commons is an activity and, if anything, it expresses relationships in society that are inseparable from relations to nature. It might be better to keep the word as a verb, rather than as a noun, a substantive' (Linebaugh 2008: 279). This is why the noun 'commons' has been expanded into the continuous verb 'commoning', to denote the continuous making and remaking of the commons through shared practice. In this way, the commons is not a static community that exists *a priori* or a society to come *a posteriori* but something that is only ever constituted through acting and doing in common. At the heart of this relational, situated interdependence of humans and non-humans is not an impoverished world of 'niggardly nature', nor an infinitely malleable world of 'techno-nature', but a more-than-human commons that navigates between limits and possibilities as they arise (Bresnihan, 2015). Nor is the more-than-human commons a pre-modern ideal that has been lost or marginalized. It arises wherever there is an immediate and intimate understanding that the world is shared, that human and non-human life are interdependent. This not an ideal norm but a materially and socially constituted reality that has been documented in many different settings (Linebaugh 2008; Scott 1990).

The problem is that while there has been much work by critical political ecologists analysing and unpicking the relations and ecologies of capitalism (Moore 2014a, 2014b), there has not been so much work examining the social relations of the commons and the everyday practices that maintain these relations (Blomley 2008; Bresnihan and Byrne 2014). Referring to the 'silence of the commons', Ivan Illich writes: 'the law establishing the commons was unwritten, not only because people did not care to write it down, but because what it protected was a reality much too complex to fit into paragraphs' (Illich 1983: 6). This is not just an empirical observation but an epistemological and methodological problem: '[t]he analytical absence of the commons from our mental maps constitutes an analytical failure' (Blomley 2008: 322).[3] While the term 'commons' has received much attention in recent years from academics, activists and policy makers, it is far from clear what it consists of or how we are supposed to identify and describe it when the intellectual and analytic tools available are so insufficient – unsurprising when they are largely inherited from an epistemology and aesthetic tradition that is literally unable to see these worlds. As Rowe rightly points out, '[b]efore we can reclaim the commons we have to remember how to see it' (Rowe 2001).

However, there are new fields of research that can help us to decipher what is going on in the more-than-human commons. These include the work of anthropologists examining indigenous cosmologies and relations with nature and territory (De la Cadena 2010; Escobar 1999; Rose 2004; Viveiros Castro 1998), as well as post-humanist and vital materialist theory (Barad 2003; Bennett 2010; De la Bellacasa 2010, 2012; Papadopoulos 2010, 2010a) that help to shift the methodological and epistemological lens away from subjects and objects to the *relata*, the relations that constitute our world (Barad 2003). These rich literatures can help us disrupt the liberal humanist epistemologies that both individualize and place humans at the centre of world-making processes. In terms of the more-than-human commons this also means making an intellectual leap into contexts where social and material resources are already immediately and intimately shared between humans and non-humans.

In this chapter, I want to tie together some insights from this diverse literature with my own empirical research on everyday practices of fishermen in Castletownbere, a commercial fishing port in the south-west of Ireland. I spent sixteen months living there and working on several boats as part of my doctoral research on the fisheries. While my research focussed on new forms of enclosure and governance in the fisheries, my everyday experiences and interactions with the place and the people who lived there revealed a 'world of fine difference' that was not easily described, let alone conceptualized. What was clear was that the social relations, knowledge and (re)productive practices that fishermen were part of constituted something different to the capitalist and biopolitical rationalities they were also caught up in.

The care of the commons

When I was living in Castletownbere, a commercial fishing port in the south-west of Ireland, a state-led project to introduce community-managed lobster fisheries around the coast was in the process of being implemented. It had been in the pipeline for over ten years as the state sought to move the lobster fisheries, and inshore fisheries in general,[4] from being an 'unregulated' fishery to being a regulated fishery where fishermen were to be issued with authorizations to fish within certain territorial zones for certain species of fish. The need for some form of regulated access had been identified since the late 1990s, when it was clear that lobster stocks were dwindling as a result of pressure from overfishing. The plan has now run in to various obstacles, most significantly a lack of support from many of the lobster fishermen themselves.

While the reluctance of fishermen to support the proposed scheme is understood by the fisheries managers involved as evidence of the 'short-sightedness' of fishermen when it comes to conserving fish stocks, a different perspective emerges when you talk to a fisherman like Joe, a young, full-time lobster fisherman who had been involved in the initial consultations on the management of the inshore fisheries. He understood fully the need for regulation but was wary of the proposed authorization system because it failed to distinguish between different kinds of fishermen. Granting an authorization to everyone who currently fished for lobsters appeared to draw an arbitrary boundary in the present, excluding considerations of the past and the future that were not reducible to bio-economic conservation targets. These concerns can help us to understand what is at stake in the more-than-human commons and what is meant by commoning. Understood as an ongoing activity, decisions over who should be included or excluded from the commons refers to how a person relates to and participates in the making of the commons, rather than being a discrete right vested in a person.[5]

For full-time fishermen like Joe, the fact that anyone with a boat could effectively fish for lobsters in the summer was a real problem. He was referring to a decision made by the Irish government in the early 2000s to grant all non-licensed fishing boats around the coast a free potting license. The reason behind this was that the state wanted to identify and include all those who were active in the inshore fisheries before devising a management plan regulating access. The problem for Joe was that these part-time fishermen didn't know much about the fishery or rely on it for their livelihoods; they could opportunistically catch fish without having to consider the past or the future.

> They could all now fish for free, and many of them had other jobs, fully pensionable jobs as teachers, gardai, civil servants, and fished 100 or so pots in the summer for beer money or whatever. They didn't have many overheads, most of them were in small punts and could handle ten pots on a string, ten strings, one string an evening. *They don't care about the stock.* It is them who hammer the stock because it doesn't matter what happened the next year, they don't rely on it [pointing out along the rocks]. People have been known to leave their pots there for a month or so, too lazy to collect them, or hoping to just catch a load in one go.
>
> (Joe)

Significantly, Joe defines these part-time fishermen more by what they don't do than what they do. Joe and other lobster fishermen I met and fished with were aware of the need to look after the lobster stocks, to ensure that they come back year after year. This is not just about limiting the number of lobsters you take, but which types of lobster you take, how often you leave your pots down, for how long and how many. The voluntary conservation practice of protecting selected mature, female lobsters by 'V-notching' their tail fan, for example, is about recognizing the importance of the lobster's reproductive cycle to the continuing (re)productivity of the lobster fishery. 'V-notching' is a practice that involves marking mature, female lobsters with a small 'V' notch in the tail. If a lobster is caught with a notch, fishermen are encouraged to return it to the sea. Fishermen are also encouraged to return berried (carrying eggs) lobsters to the water. Holding a fertile female lobster up to me, Tom, a long-time fisherman said: 'a fisherman killing a berried hen is like a farmer killing his cow when she is in calf'. The ongoing reproduction of the lobster is something that lobster fishermen cannot afford to ignore; they rely on the lobsters and thus can't mistreat them by leaving their pots down for more than a few days, or keeping back small lobsters or fertile female lobsters. This 'common-sense' understanding is not based on an abstract conservation ethic but emerges from an immediate, material constraint imposed on their activity by the fact that they are dependent on the lobsters.

Fishermen like Joe are not just aware of and reliant on the activity of lobsters. They are also aware that they share the fishery with other fishermen, fishermen whom they know and see everyday. In the south-west of Ireland, all the full-time fishermen are known to each other. When a small punt appeared some distance away, obviously fishing pots in close to some cliffs, I asked Joe if he was one of the part-time fishermen he was complaining about. Joe immediately knew who it was – a semi-retired fisherman – and said he wasn't the problem because he knew how to take care of the fish. He knows this because he lives and works beside this fisherman. This isn't mysterious; it is the kind of knowledge that one gathers from everyday interactions and exchanges on land or at sea. This social, situated knowledge means that formal rights of access, such as authorizations, have not been required until recently. Instead, informal norms and practices exist concerning when and where fishermen can drop their pots and nets.[6] In Kerry, for example, where there are even fewer lobster fishermen than in Castletownbere, the only practice they observe is a first-come-first-drop rule. Around other parts of the West Cork coast where the fishing is busier I twice encountered situations where a fisherman's strings had been cut. This means that the buoy attached to the pots that lie along the sea-floor is cut loose so that retrieving the pots becomes very hard. On both occasions the fishermen agreed it was their own fault, even though they stood to lose several thousand euro-worth of gear. This was because the pots had been dropped somewhere they shouldn't have been. This response springs from an immediate and intimate understanding that the fisheries are shared and thus individuals can not just do what they want.

Where the more-than-human commons departs from other interpretations is in recognizing how the starting point is not an individual subject separated from other people and the world around them, but a relational subject who is always already caught up in a world that is intimately shared. This understanding is not based on an ideal but on the materially and socially constituted relations and practices that tie humans and non-humans together within a particular collective or territory. If we talk of 'use-rights' in the commons, then these must be contingent on ongoing participation in the production and care of the commons understood as the entire collective of humans, animals, artefacts, elements that are necessary to maintain life processes. This meaning can already be found in the roots of the word 'commons': 'com' (together) and 'munis' (under obligation). First, this tells us that the commons is produced together, reflecting our interdependence, the assumption that our world is already shared. Second, and arising from this, the obligation that such interdependence demands of us. The commons is not a 'thing' that we have access to because we hold a title deed or authorization, but something that is ours because we produce and care for it, because we common.[7]

The concept of 'care' is helpful in trying to articulate the ethos underlying these situated practices of commoning. Care denotes the immediate interdependency of human and non-human life; it expresses relations between particular communities of humans and non-humans that are not fixed or prescribed in advance but are worked out according to and across a variety of different needs and interests. Maria Puig de la Bellacasa quotes Tronto's general definition of 'care' as 'everything that we do to maintain, continue and repair "our world" so that we can live in it as well as possible. That world includes our bodies, ourselves, and our environment, all that we seek to interweave in a complex, life sustaining web' (quoted in de la Bellacasa 2010). The idea that care holds the world together through an interweaving of all aspects of our lives returns us to the interconnectedness and interdependency of human and non-human life. Such an understanding brings us beyond liberal notions of ethics that focus on the individual and 'self-care', to the individual as part of a human and non-human collective that must be nourished on an ongoing basis. As Deborah Bird Rose writes 'ethics are situated in bodies and in time and in place' (Rose 2004: 8). Care thus points to the wide scope of attention required to ensure an activity is done 'well', in contrast to simply measuring the output of that activity. This is where a grounded ethos of care differs from other kinds of biopolitical, output-orientated management tools being applied within resource management.

De la Bellacasa uses the example of permaculture to help communicate the fundamental differences between an ethos of care and more normative, liberal conceptions of ethics. The ethos of permaculture first of all attracts our attention to the 'invisible but indispensable labours and experiences of earth's beings and resources' (de la Bellacasa 2010: 165). Recognizing our reliance on these labours immediately decentres any sense we might have of our own agency, while at the same time placing us into a relationship with nature that is not abstract but always material and situated. It requires ongoing action and care, not as an additional component or prescription on our 'normal' activity but as an integral part of how we do things. This involves knowledge practices that are distributed across the more-than-human needs and agencies that underlie the (re)production of the commons.[8]

These ways of doing care are grounded in an ethos that respects and understands limits at the same time as it respects and understands potential and possibility. Navigating these changeable parameters is the practical, situated 'doing' of care work, which is to say, commoning.[9] This is a constantly negotiated series of socio-natural relationships as Mabel McKay, a Powo healer, observes: 'when people don't use the plants, they get scarce. You must use them so they will come up again. All plants are like that. If they're not gathered from, or talked to and cared about, they'll die' (quoted in de la Bellacasa 2010: 161). While the relational, situated nature of commoning means that liberal institutions of exclusion and ownership do not make sense, this does not mean that there are no limits on what an individual can and cannot do. It is precisely the social and material interdependency of the commons that requires practices of mutualism and reciprocity. Rather than breaking the world up into discrete parcels and units for individuals to exclusively own or manage, practices of mutualism and reciprocity are based on a recognition of limits and the need for cooperation. These limits are not abstract, quantitative figures (quotas, for example), but the concrete material limits that inhere in any ecological collective that an individual is part of and relies on.

When I learnt how to tie knots on the boats I was told that a good knot should be as easy to undo as it is strong. This suggests something of the way in which events are not predictable, that tying and untying, connecting and disconnecting, emerge in response to the limitations and opportunities presented from moment to moment; you never know when conditions will change and something will have to be let go, or something brought into play. This necessarily open and yet prescribed orientation to the unfolding of everyday natures was constantly in evidence in Castletownbere and exists wherever boundaries are conceived more as something to cross rather than as a form of exclusion. Deborah Bird Rose, for example, describes the constant negotiation and fluidity of boundaries associated with the Yolngu Aboriginal people in Australia: '[f]or

Yolngu . . . boundaries on land mark discontinuities: changes in ownership. But for Yolngu, boundaries do not exist primarily for the purpose of excluding non-owners. Rather, Yolngu use boundaries to express various categories of rights, both to users and owners' (quoted in Rose 1996: 45). Rose can explain this relationship to boundaries and ownership only by recourse to the fundamentally different ontology that underlies it. Understanding this form of organization within a liberal epistemology, 'bifurcating' as it does Humanity and Nature, subject and object, is not so much difficult as pointless. Only by assuming the perspective that immediately and intimately understands that our more-than-human world is shared and interdependent can we recognize the material necessity of these situated practices of care.

The circulation of the commons

We can only properly make sense of how an intimately shared world gives rise to certain limits and possibilities by expanding the territory of the commons, by placing the activity of lobster fishing, for example, within the wider social and ecological activity that operates in a place like Castletownbere. By focussing on specific resources and the people who exploit those resources we can get lost in the same bioeconomic framing that tends to dominate approaches to environmental management. In doing this we ignore the circulation of the commons, the continuous ways in which a diversity of social and material resources are mobilized through commoning.[10] From this perspective, what begins to come into view is a world that is not orientated around the production and management of scarcity, but one that relies on the circulation of surplus.

Most fishermen in Ireland are not paid a wage, rather, their income depends on what they catch. In the inshore sector most of the boats are owned and skippered by the same person. This means that they are chronically insecure and precarious because they don't make any income unless they can catch and sell their fish, a situation that pincers them between the vagaries of the sea and the market. But it also means that these individuals directly control their own means of production and labour. They are not directed or managed by a boss or the clock. It also means that they can, and must, rely on other forms of economic activity outside the wage and commodity market. It means that they must be inventive and attentive to the diversity of use-values and non-monetary exchange that exists around them. This can be described as the informal or invisible economy of the commons.

Joe makes most of his income from lobster because they are the most valuable fish, but he also targets crab with the same pots, as well as flatfish and mackerel with his nets. He values the flexibility this brings. It allows him to shift between different fishing grounds and species, depending on changes in the weather or movements of the fish. For example, lobster pots tend to be dropped closer in to shore. This means that if the weather is bad it isn't always possible to shoot for lobster or haul pots that have already been shot. If Joe knows that the weather will turn bad then he is better off shooting his pots further from the shore to target crabs. While having a boat fitted out with nets makes him less mobile in the water, potentially limiting him from accessing other fishing grounds, it opens up other possibilities in terms of what he is able to catch. Most importantly, the nets make him almost completely self-sufficient for bait (after fuel, bait is the most significant cost for crab and lobster fishermen). Nearly all the potting fishermen I met had devised ways of accessing free or cheap bait. This involved keeping the dogfish and conger eel that appeared in the pots, using by-catch from nets, or else coming to some arrangement with other fishermen they knew who went trawling. Tom, for example, got under-sized mackerel off his son's boat and kept it salted in an abandoned hold by the fish factory over the year. Tom was thus able to make a living from lobster fishing not simply by extracting and selling lobsters, but through social and material resources he was able to mobilize on land and sea through a network of relations.[11]

Maria Mies has written about the importance of this localized circulation of production and consumption as an alternative way of understanding 'waste'. While capitalist modes of production tend to separate the point of production from the point of consumption through the commodity circuit, commons-based economies incorporate the two, transforming 'waste' into something productive and reproductive (Goldstein 2013; Mies and Bennholdt-Thomsen 2001). This is because the value and wealth that circulates through the commons is not transformed into exchange-value commodities for sale or profit. 'Production processes will be oriented towards the satisfaction of needs of concrete local or regional communities and not towards the artificially created demand of an anonymous world market. In such an economy the concept of waste, for example, does not really exist,' writes Mies and Bennholdt-Thomsen (2001: 1011). The redefinition and recirculation of 'waste' is what ensures the productivity of the commons, generating a wealth of use-values rather than a limited number of commodities with market value.

While Mies focuses on the ways waste is recirculated within the commons as an immediate use-value, it is also the case that waste can be produced that does not have an immediate use. This does not mean it is discarded; things can always become useful or important by being invested somewhere else. You never know when conditions will change and something may be needed or brought into play. In this sense, it may be better to speak of the circulation of surplus rather than waste, and the different ways this surplus becomes invested in or circulated through the commons. In this the work of Bataille and his concept of the 'general economy' can be instructive (Bataille 1991). Rather than analysing economy from the perspective of production and the management of scarcity, as liberal political economy does, Bataille approaches the economy from the perspective of consumption of wealth and the management of surplus. His economic anthropology of different cultures and civilizations examines how surplus wealth is socialized through forms of ritualistic destruction, lavish consumption or accumulation and war. Bataille does not, however, examine subsistence economies where the expression 'I store my meat in the belly of my brother' carries such significance. This expression conveys the simple idea that by gifting surplus resources to a neighbour in the present, one is effectively investing in the future. In this sense, where the care-practices involved in commoning are more about recognizing material limits that inhere in the commons, the practices involved in circulating surplus relate to the production and store of wealth. As Linebaugh writes: '[t]he commons is not a natural resource exclusive of human relations with it. *Like language itself, the commons increases in wealth by use*' (Linebaugh 2012: 21; my italics).

The first job I got in Castletownbere was harvesting mussels on a mussel farm for a few weeks. At the end of each day, the mussel farmer, Frank, would leave the harvested mussels at a small pier for collection by the wholesaler, who was also a neighbour of Frank's. One day Frank put the harvested mussels on his trailer and dropped them over to the bigger pier so they would be easier for the wholesaler to pick up. When I asked him why he did it he said: 'You just never know, I might need his help some time if I break down or whatever.' In a similar way, Frank always collected the mussels that had missed the bags and were scattered on the deck of the boat. By law these mussels were supposed to be discarded at sea for health reasons, but Frank always collected them and put them into a net bag which he hung from the side of the pier. He left them there for neighbours and friends and told me if I ever wanted mussels I could come by and just take as many as I wanted.

These small acts of generosity may appear trivial but they were ever-present in Castletownbere. While they were often incidental and taken for granted, they constituted an invisible network of favours and gifts that operated like a reserve to be drawn on at any moment of need or crisis. Nor were these gifts of time, labour and resources, calculated in the terms of straightforward utilitarianism; they were not the actions of individuals working out exactly what was in his or her own self-interest. They were the actions of people who knew intimately and immediately that they were part of a wider collective on which they relied. The refrain 'you never know' corresponds to an immediate awareness of how unpredictable things are and how the continuing support of others allows you to deal with this unpredictability. This support does not come naturally, nor does it rely on fixed actions or a form of paid membership. Taking part in this wider collective is an ongoing, everyday social practice that almost recedes out of sight when it is reproduced through such minor acts of gift giving.

The role of these social relations of reciprocity, mutualism and gift giving in the making and remaking of the commons has been observed in many different contexts (Graeber 2011; Scott 1977).

> What in the Andes is called *ajnji* is a form of reciprocal labour that is the weaving of the social fabric of a community (or *Ayullo* in the Andes) through circuits of reciprocity, and it is based on principles of often implicit and not-announced or bargained equality matching between individuals or community: today you do this for me and tomorrow I'll do this for you: a kind of circular 'gift economy' as discussed by Mauss
>
> (De Angelis 2007: 176).

These relations of reciprocity do not just enable a form of non-market (re)production by 'saving on costs'. As with the situated practice of care, reciprocity within the commons is an ongoing form of exchange and interaction that can not easily be separated into a discrete 'economic' sphere. De Angelis makes the point that such sharing is often characterized by conviviality, but, equally, it can give rise to arguments and conflict. This pushes us into the middle of the drama, the 'theater within which the life of the community is enacted and made evident' (Hyde 2011: 31), where the messy give and take of commoning unfolds in response to practical needs and interactions. It is the practical, situated nature of commoning that makes it hard to see: a social and material backdrop that is usually recognized and valued only after it has disappeared. This invisibility is the real tragedy of the commons and it raises questions about how such practices of commoning can translate into a form of politics capable of challenging the hegemony of biopolitical and capitalist enclosure.

Enclosure and the manifold commons

While Joe complained about the part-time fishermen taking lobsters during the summer holidays, he was clear about where the real problem lay: the price of fish. Fishing wasn't viable anymore. It didn't matter how healthy the stock was, how well regulated, if a fisherman couldn't survive off his income. At that time the price of lobster was €9.50/kg instead of €13/kg the year before, and brown crab was €1.05/kg rather than €3 or €4/kg. While prices fluctuate and may well return, the reality is that the cost of fuel is going up, more people are moving into the inshore sector as they are forced out of the competitive offshore fisheries and retailers and processors are under no pressure to pass on any extra money if prices do improve. Because of these pressures fishermen are often being forced to overfish or, more specifically, to engage in practices that undermine the commons.

The day I went potting with Joe he had been told by the processing factory that it wasn't accepting the bodies of crabs as it couldn't sell them. The processing factory had been in decline and Joe was worried he'd have nowhere to sell his catch – or that he'd have to sell to a processing company further down the coast, which would involve using more fuel. That day he (and I) had no choice but to rip the claws off all the crabs he caught – maybe two to three hundred. It is illegal to rip both claws off the crab because they have no way of hunting, essentially ensuring that the crab has no way to survive. Joe was also aware of how painful it must be for the animal. Joe knew all this but had no choice. Not only did he run the risk of a fine and a criminal record if he was caught, but he also thought it was a 'terrible waste of good meat'.

While the 'exogenous violence of the market' (Caffentzis 2010) penetrates deeper into the everyday relations of production of the inshore fisheries, inshore fishermen also face 'creeping enclosure' from environmental regulations orientated around bio-economic metrics of sustainability and conservation (Murray et al. 2010). As with the plan for community-managed lobster fisheries, such regulatory frameworks do not distinguish between different ways of fishing, different relations of (re)production, that continue to exist within the inshore fisheries.[12] But this raises a recurring problem: the relations and practices that sustain the commons are rarely visible until they disappear. This is the real meaning of the tragedy of the commons. While Joe and other fishermen I worked with were angry at what was happening to the fisheries, and sharp to see and speak about where the problems lay, they were not moved to do anything about it. They were caught between the need to make a living, to extract increasing amounts of fish, to repay debts and the less easily articulable values and meanings that inhere in the place they live in and love. The problem, as Deborah Bird Rose suggests, is that our capacity for seeing and articulating the less articulable world of meaning and value has been eroded at the same time as capitalist relations of production and biopolitical rationalities have extended: 'appropriation deprives people of their power to be present to others in their own history and knowledge' (Rose 2004: 200). This forces us to ask the question, especially now, of what the connection is, or can be, between everyday practices of commoning and the formation of alternative claims to territory and forms of life against capitalist and biopolitical forms of enclosure.[13]

Reflecting on the upsurge of indigenous politics in Latin America over recent decades, Marisol de la Cadena writes of how the disagreements underlying these new political antagonisms are not understandable within the familiar terms of Western political philosophy and political economy. While these conflicts have arisen in response to capitalist accumulation processes, the indigenous response has not tended to fit within Western ontologies that identify humans as political subjects and non-humans as apolitical (Latour 1993). 'Digging a mountain to open a mine, drilling into the subsoil to find oil, and razing trees for timber may produce more than sheer environmental damage or economic growth,' she writes. 'These activities may translate into the violation of networks of emplacement that make life locally possible – and even into the destruction of place' (De la Cadena 2010: 357). While De la Cadena identifies the importance of place in these struggles, this should not be read simply as a defence of the local community or resources against the incursions of global power and capital. The concept and practice of place within Andean cosmopolitics signifies a fundamentally different relationship between humans and non-humans that amounts to an alternative form of life.[14] These movements have thus sought to bring 'earth-beings' and 'earth-practices' into the political sphere, to insist that relational ontologies be recognized rather than separated and rendered manageable within existing discursive frameworks and political practice. What is at stake in these new indigenous movements is not just access to resources or the value generated through their extraction but a radical challenge to the epistemological dichotomy of Nature and Humanity, and thus a disagreement over what ecological arrangements count and who is capable of speaking for them.

The sense in which Andean cosmopolitics represents not just a different claim to resources but a different world of relations between humans and non-humans, a different form of life, resonates with a long history of struggles against what Marisol de la Cadena calls the 'singular biopolitics of improvement' (De la Cadena 2010: 346).[15] While the commons has a history, and a historical continuity, it is not a pre-modern form of life or cosmology that is lost forever. Understood as a more-than-human commons, it not only challenges capitalist modes of production but humanist (dualist) assumptions of agency and social change. The existence and appearance of forms of (re)production grounded in an immediate and intimate sense that the world is shared between humans and non-humans runs alongside and underneath dominant narratives of modernity and progress; it provides us with a counter-modern reference, a 'different praxis and management of the material dimension of life – in other words, an entirely different "economics", or way of ensuring the satisfactions of our material wants and needs', as Freya Mathews puts it (1999: 130), that can always 'arise unexpectedly in relationships among peoples and between people and place' (Rose 2004: 6).[16]

Rather than too readily short-circuiting the emergence of the more-than-human commons as a new site of political struggle through an appeal to existing forms of political representation, it is important to pay attention to the novel material and discursive practices that are opening around the commons, producing worlds in commons: shared worlds that have an immanent power, an internal logic, that escape the value and knowledge practices of capitalist accumulation and biopolitical control. From this perspective, the construction of a politics of the common(s) is going to emerge only through a process of collective subjectivization, the making of new subjects and 'values that are grounded in material practices for the reproduction of life and its needs' (De Angelis 2007 32).[17] As Maria Puig de la Bellacasa writes, we need to find alternative ways of world-making that cultivate 'power with' rather than 'power over' the more-than-human world (Puig de la Bellacasa 2010). Cultivating different ways of producing and sustaining life in common does not mean a 'return' to a 'pure' pre-capitalist time, nor even a valorization of subsistence while wealth continues to be accumulated by global elites. First and foremost, it is about a creative elaboration, intensification and expansion of our life-needs in a process of collective (human and non-human) development and experiment (Dyer-Witheford 2006).

Notes

1 The most well-known of these metaphors is Garret Hardin's 'tragedy of the commons': 'Ruin is the destination toward which all men rush, each pursuing his own best interest in a society that believes in the freedom of the commons' (Hardin 1968: 1244).
2 And just as women in Europe in the seventeenth and eighteenth centuries were to the forefront of struggles for access to common lands that were being enclosed, so now in parts of the global South women are doing the same because they are often the primary producers of food, as well as the harvesters and gatherers of various other natural resources necessary for energy and healthcare. Maria Mies, who often relies on stories to articulate the entwining of women, nature and peasant within her 'subsistence perspective', narrates the story of her own mother who, in the face of general shortages of food, refrained from slaughtering their last pig at the end of World War II because 'life must go on': she knew the pig would have piglets (Mies and Bennholdt-Thomsen 1999).

3 In historical terms, Peter Linebaugh describes how evidence of the commons will often appear anecdotal or as folklore or as 'crime', just a small story, a minor transgression; evidence of commons may appear incidentally to some other, major theme; evidence of customary commons may appear particular to locale or craft, and belonging thus to trade or local histories, not 'grand narratives' (Linebaugh 2008). Echoing this, Neeson writes: '[i]f the measurement of crime has its dark figure of criminal acts not reported, common right has its dark figure too – of practice not recorded' (Neeson 1996: 79–80). Neeson shows us that the commons were not reducible to archaic customs or laws: this was not their only, or most significant, form. Rather, the commons were organized through everyday social practices that negotiated the varying needs, possibilities and limits that emerged as people directly interacted with, produced and cared for the resources they relied on.

4 The inshore fisheries encompasses a geographical area that extends six nautical miles from the national coast. Inshore boats are usually no longer than 10m and crewed by a maximum of two people – the skipper/owner and a crewman. They generally head out early in the morning and return in the early evening. Fishing is mixed and varied, depending on the season, with the most profitable species being shellfish such as lobster, crabs and prawns supplemented with small catches of mixed white fish and mackerel. The most valuable inshore species in terms of earnings is the lobster.

5 A direct parallel can be made between Jeanette Neeson's observation about claims for compensation in the eighteenth century at the time of enclosure and claims for authorizations to fish lobsters in the Irish inshore fisheries. Just as lobster fishermen must prove that they have a track record of fishing for lobster, Neeson writes that communities who had used common lands were required to provide material proof or evidence that they had a customary right in order to be compensated for their loss of access to the land. This evidence was often made up in order to accord with the legal requirements, when in actual fact the right of access was enshrined in the more complex pattern of customary practices that were performed, not written on paper.

6 Ostrom quotes a similar example of how fishing is policed in Davis. Here, a fisherman sees another setting his gear near or even over his own gear. He calls the other fisherman across the radio and complains. Someone else on the radio agrees and the 'transgressor' puts his gear somewhere else (Ostrom 2008).

7 Jeanette Neeson, and other social historians, have described something similar when referring to the existence of 'customary rights' governing the use of land, rivers, turfmoors and forest. These were the rights of pasturage, piscary, turbary and estovers. 'Custom' is another word for practice or usage, and such rights were fleshed out through elaborate practices of 'stinting'. Stinting was not just about access but about use, about recognizing and acting in accordance with particular limits – limits that were not 'absolute', but were situated, reflecting the social and ecological knowledge that was required in order to take care of the commons. For example, when to harvest, how to harvest, how much and by whom. Stinting was a practical, social regime for sharing the product of the commons – recognizing the needs and participation of humans and non-humans in this productive process.

8 As Justo Oxa, an indigenous school teacher from Peru, writes, 'respect and care are a fundamental part of life in the Andes; *they are not a concept or an explanation*. To care and be respectful means to want to be nurtured and nurture others, and this implies not only humans but all world beings' (quoted in De la Cadena 2010: 354; my italics). He is referring here to the Andean concept of *uyway* : '[e]mbedded in everyday practices, uyway refers to mutual relations of care among humans and also with other-than-human beings' (quoted in De la Cadena 2010: 354).

9 The philosopher and biologist Francisco Varela refers to something similar in the work of the Chinese philosopher Mencius. Mencius articulated the concept of 'wu-wei',

which translates as 'not-doing'. This ethos suggests that there should be no ground for action on the basis that we are at our most attentive when we need to extend into the world, when the world is not already predefined. The example he gives is when seeing a boy on the edge of a cliff about to fall we are not moved to help because of a desire for praise or out of a dislike for the cry of the child. The idea of 'wu-wei' removes intentionality, pointing instead to the way in which events call on actions which are not merely responses to stimuli, nor reasoned decisions, but something else. While there may be a need for normative rules in society, Varela argues, these rules need to be informed by the wisdom that enables them to be dissolved in the demands of responsivity to the immediacy of lived situations.

10 Tim Ingold uses the term 'taskscape' to describe the way places and people are co-produced through activities. He writes: '[i]t does not begin here (with a pre-conceived image) and end there (with a finished artefact), but is continuously going on' (Ingold 2000: 205).

11 This reliance on a variety of resources that may have multiple uses is reminiscent of Starhawk's evocation of the historic commons. 'Enclosed land, instead of serving multiple needs and purposes, served only one. When a forest was cut down and enclosed for grazing land, it no longer provided wood for fuel and building, acorns for pigs, a habitat for wild game, a source of healing herbs, or shelter for those who were driven to live outside the confines of town and village. When a fen was drained to provide farmland, it no longer provided a resting place or nesting sites for migratory birds, or a source of fish for the poor' (Starhawk 1982: 189).

12 Another example is a ban on discards has recently been introduced by the European Union and applies to all fisheries. While this is intended to prevent the wasteful practices of discarding that take place across intensive, large-scale offshore fishing, it technically means that the practice of keeping by-catch as free bait for lobster and crab pots is illegal. Without this free source of bait, fishermen will be forced to adopt more specialized (and intensive) modes of fishing, like the seafood company, or even start working for the company directly.

13 In a similar sense, Lewis Hyde describes the old custom of 'beating the bounds', a festive and potentially antagonistic ritual activity carried out every year by commoners in England. This involved walking together around the bounds of the territory of the commons, breaking down any walls and ditches that might have been erected in the meantime. He asks: '[w]hat would be the [contemporary] equivalent of an annual perambulation with cakes and beer, the inhabitants armed with axes and crowbars to act on the ancient rule that citizens may tear down enclosures when they arise?' (Hyde 2011: 242).

14 This concept is informed by the work of Nick Dyer-Witheford on the radical possibilities of Marx's category of 'species being' (Dyer-Witheford 2006); Dimitri Papadopoulos, who has argued for a radically materialist politics of 'worlding', 'an opening to material processes and practices as a possibility for crafting – literally – common, alternative forms of life' (Papadopoulos 2012: 2); and Maria Puig de la Bellacasa, who offers an analysis of situated, relational ethics in the production of nature-cultures as a form of 'alter-biopolitics' (de la Bellacasa 2010).

15 Foucault describes how the rise of biopolitics in the eighteenth century precipitated new forms of opposition demanding recognition and rights in the name of the body and of life. 'Against this power [biopower]', he writes, 'the forces that resisted relied for support on the very thing it invested, that is, on life and man as a living being . . . [W]hat was demanded and what served as an objective was life, understood as the basic needs, man's concrete essence, the realization of his potential, a plenitude of the possible' (Foucault 1998: 144–145). Marzac confirms this when he describes how in the eighteenth century '[t]he land was represented as a raw material in need of "improvement" and "cultivation", and no longer as an entity that gave life to "inhabitants", a key term used by those who resisted enclosures' (Marzac 2011: 83).

16 This echoes Nicholas Blomley's analysis of an anti-enclosure struggle in Vancouver (Blomley 2008). While the urban architecture which was going to be commercialized and developed was considered 'abandoned', it was layered over with years of historical memory and experience, experiences which continued to materialize through the everyday lives of those who lived near the building. The campaign to resist the 'enclosure' of the building enacted these qualities by using and re-inhabiting the space. This was not just an identifiable 'community' fighting for an urban 'resource' but, rather, a material politics which demonstrated and referred to the manifold relations and values which were constituted over time through the use of the space. Blomley, trying to find some way of expressing this common right or value, settles on the 'right to not-be-excluded', that is, a negative right of the commons which cannot adequately represent its concerns and experiences within existing categories (of ownership) but nonetheless can still resist the claim of any individual or collective to exclude others.

17 De la Cadena adopts the idea of 'slowing down reasoning' from Isabelle Stengers: '[e]merging through a deep, expansive, and simultaneous crisis of colonialism and neo-liberalism – converging in its ecological, economic, and political fronts – the public presence of unusual actors in politics is at least thought provoking. It may represent an epistemic occasion to "slow down reasoning" . . . and, rather than asserting, adopt an intellectual attitude that proposes and thus creates possibilities for new interpretations' (De la Cadena 2010: 336).

References

Acheson, J. M. and McCay, B. J. (Eds.) (1990). *The Question of the Commons. The Culture and Ecology of Communal Resources*. University of Arizona Press, Tucson.

Agrawal, A. (2003). Sustainable governance of common-pool resources: Context, methods, and politics. *Annual Review of Anthropology*, 32, 243–262.

Barad, K. (2003). Posthumanist performativity: Toward an understanding of how matter comes to matter. *Signs*, 28(3), 801–831.

Barrell, J. (2010). *The Idea of Landscape and the Sense of Place 1730–1840. An Approach to the Poetry of John Clare*. Cambridge University Press, Cambridge.

Bataille, G. (1991). *Accursed Share: Volume 1: Consumption*. Zone Books, Cambridge, MA.

Bauwens, M. (2005). P2P and human evolution: Peer to peer as the premise of a new mode of civilization, http://library.uniteddiversity.coop/Money_and_Economics/P2P_essay.pdf.

Bennett, J. (2010). *Vibrant Matter: A Political Ecology of Things*. Duke University Press, Durham and London.

Blomley, N. (2008). Enclosure, common right and the property of the poor. *Social & Legal Studies*, 17, 311–331.

Bresnihan, P. (2015). *Neoliberalism, Nature and the Commons*. University of Nebraska Press, Lincoln, NE.

Bresnihan, P. and Byrne, M. (2015). Escape into the city: everyday practices of commoning and the production of urban space in Dublin. *Antipode* 47(1), 36–54.

Brody, H. (2002). *The Other Side of Eden: Hunters, Farmers, and the Shaping of the World*. Macmillan, London.

Caffentzis, G. (2010). The future of 'The Commons': Neoliberalism's 'Plan B' or the original disaccumulation of capital? *New Formations*, 69(1), 23–41.

Cooper, M. (2008). *Life as Surplus: Biotechnology and Capitalism in the Neoliberal Era*. University of Washington Press, Seattle.

De Angelis, M. (2007). *The Beginning of History: Value Struggles and Global Capital.* Pluto Press, London.

De la Bellacasa, M. P. (2010). Ethical doings in nature cultures. *Ethics, Place and Environment*, 13, 151–169.

De la Bellacasa, M. P. (2012). 'Nothing comes without its world': Thinking with care. *The Sociological Review*, 60(2), 197–216.

De la Cadena, M. (2010). Indigenous cosmopolitics in the Andes: Conceptual reflections beyond 'Politics'. *Cultural Anthropology*, 25(2), 334–370.

Dietz, T., Ostrom, E., and Stern, P. C. (2003). The struggle to govern the commons. *Science*, 302(5652), 1907–1912.

Dyer-Witheford, N. (2006). Species-being and the new commonism: Notes on an interrupted cycle of struggles. *The Commoner*, 11, 15–32.

Escobar, A. (1995). *Encountering Development: The Making and Unmaking of the Third World.* Princeton University Press, Princeton.

Escobar, A. (1999). After nature: Steps to an antiessentialist political ecology 1. *Current Anthropology* 40(1), 1–30.

Federici, S. (2001). Feminism and the politics of the commons. *The Commoner.* http://andandand.org/pdf/federici_feminism_politics_commons.pdf (last accessed 29 June 2013)

Federici, S. (2012). *Revolution at Point Zero: Housework, Reproduction, and Feminist Struggle.* PM Press, Oakland, CA.

Foucault, M. (1998). *The History of Sexuality, Vol. 1: The Will to Knowledge.* Penguin Books, London.

Goldman, M. (2004). Imperial science, imperial nature: Environmental knowledge for the World (Bank). In S Jasanoff and M. L. Martello (Eds) *Earthly Politics. Local and Global in Environmental Governance.* The MIT Press, Cambridge, MA.

Goldstein, J. (2013). Terra Economica: Waste and the production of enclosed nature. *Antipode*, 45(2), 357–375.

Graeber, D. (2011). *Debt: The First 5,000 Years.* Melville House, London, UK.

Hanna, S. S., Folke, C., Maler, K-J. (Eds.) (1996). *Rights to Nature. Ecological, Economic, Cultural and Political Principles of Institutions for the Environment.* Island Press, Washington, DC.

Hardin, G. (1968). The tragedy of the commons. *Science*, 162, 1243–1248.

Hardt, M. (2010). Two faces of apocalypse: A letter from Copenhagen. *Polygraph*, 22, 265–74.

Hardt, M. and Negri, A. (2009). *Commonwealth.* Harvard University Press, Cambridge, MA.

Hyde, L. (2010). *Common as Air: Revolution, Art, and Ownership.* Macmillan, London.

Illich, I. (1983). Silence is a commons. *CoEvolution Quarterly* 40, 5–9

Ingold, T. (2000). *Perception of the Environment: Essays on Livelihood, Dwelling and Skill.* Routledge, London.

Latour, B. (1993). *We Have Never Been Modern.* Harvester Wheatsheaf, Herefordshire.

Leach, M. (2008). Pathways to sustainability in the forest? Misunderstood dynamics and the negotiation of knowledge, power, and policy. *Environment and Planning A*, 40(8), 1783–1795.

Li, T. M. (2006). Neo-liberal strategies of government through community: The social development program of the World Bank in Indonesia. Institute for International Law and Justice. https://tspace.library.utoronto.ca/bitstream/1807/67415/1/2006-2-GAL-Li-final-web.pdf.

Linebaugh, P. (2008). *The Magna Carta Manifesto*. University of California Press, California.

Linebaugh, P. (2011). Enclosure from the bottom up. *Radical History Review*, 108, 11–27.

Linebaugh, P. (2012). *Ned Ludd and Queen Mab. Machine-Breaking, Romanticism, and the Several Commons of 1811–12*. Retort Pamphlet Series. PM Press, Oakland, California.

McCann, A. (2004). Enclosure within and without the commons. http://dlc.dlib.indiana.edu/dlc/bitstream/handle/10535/892/mccannoaxaca.pdf?sequence=1 (last accessed 2nd October 2014).

Mansfield, B. (2004). Neoliberalism in the oceans: 'Rationalisation', property rights, and the commons question. *Geoforum*, 35, 313–326.

Marzac, R. (2011). Energy security: The planetary fulfilment of the enclosure movement. *Radical History Review*, 109, 83–100.

Mathews, F. (1999). Letting the world grow old: An ethos of countermodernity 1. *Worldviews: Global Religions, Culture, and Ecology*, 3(2), 119–137.

Mies, M. (1998). *Patriarchy and Accumulation on a World Scale: Women in the International Division of Labour*. Palgrave Macmillan, London.

Mies, M., and Bennholdt-Thomsen, V. (1999). *The Subsistence Perspective: Beyond the Globalised Economy*. Spinifex Press, Melbourne, Australia.

Mies, M., and Bennholdt-Thomsen, V. (2001). Defending, reclaiming and reinventing the commons. *Canadian Journal of Development Studies/Revue canadienne d'études du développement*, 22(4), 997–1023.

Moore, J. W. (2014a). The Capitalocene, Part I: On the nature and origins of our ecological crisis. Unpublished paper, Fernand Braudel Center, Binghamton University.

Moore, J. W. (2014b). The Capitalocene, Part II: Abstract social nature and the limits to capital.

Murray, G., Johnson, T., McCay, B. J., St. Martin, K., Takahashi, S. (2010). Cumulative effects, creeping enclosure, and the marine commons of New Jersey. International Journal of the Commons, 4(1), 367–389.

Neeson, J. M. (1996). *Commoners: Common Right, Enclosure and Social Change in England, 1700–1820*. Cambridge University Press, Cambridge.

Ostrom, E. (2000). Collective action and the evolution of social norms. *Journal of Economic Perspectives*, 14(3), 137–158.

Ostrom, E. (2008). *Governing the Commons: The Evolution of Institutions for Collective Action*. Cambridge University Press, Cambridge.

Ostrom E. and Schlager, E. (1992). Property-rights regimes and natural resources: A conceptual analysis. *Land Economics*, 68(3), 249–262.

Papadopoulos, D. (2010). Insurgent posthumanism. *Ephemera*, 10, 134–151.

Papadopoulos, D. (2010a). Alter-ontologies: Towards a constituent politics in technoscience. *Social Studies of Science*, 41, 177–201.

Rose, D. B. (1996). *Nourishing Terrains: Australian Aboriginal Views of Landscape and Wilderness*. Australian Heritage Commission, Canberra, Australia.

Rose, D. B. (2004). *Reports from a Wild Country: Ethics for Decolonisation*. University of New South Wales Press, Sydney, Australia.

Rose, N. (2006). Governing 'advanced' liberal democracies. In A. Gupta and a. Sharma (Eds) *The Anthropology of the State: A Reader*. Blackwell Publishing, Oxford

Rowe, J. (2001). The hidden commons. http://www.yesmagazine.org/article.asp?ID=443.

Scott, J. C. (1977). *The Moral Economy of the Peasant: Rebellion and Subsistence in Southeast Asia*. Yale University Press, New Haven, CT.

Scott, J. C. (1990). *Domination and the Arts of Resistance: Hidden Transcripts*. Yale University Press, New Haven, CT.

Shiva, V. (2010). Resources. In Sachs, W. (Ed.) *The Development Dictionary*. Zed Books, London, 228–243.

Starhawk (1982). Appendix A. *The Burning Times: Notes on a Crucial Period of History, Dreaming the Dark*. Beacon Press Book, Boston, MA.

St Martin, K. (2009). Toward a cartography of the commons: Constituting the political and economic possibilities of place. *Professional Geographer*, 61, 493–507.

Thompson, E.P. (1993). *Customs in Common. Studies in Traditional Popular Culture*. The New Press, New York.

Viveiros de Castro, E. (1998). Cosmological deixis and Amerindian perspectivism. *The Journal of the Royal Anthropological Institute*, 4(3), 469–488.

5 'Where's the trick?'

Practices of commoning across a reclaimed shop front

Mara Ferreri

> We, the OffMarket collective, take disused buildings and turn them into open resources. We have just moved in to [address in North-East London]. This space has been empty for more than a year now. Yes, we occupy buildings that don't 'belong' to us. [. . .] There are lots of ideas about what to do with this space! One is to initially try and link to local and broader struggles that are happening in Hackney around issues like the coming cuts, housing, employment, gentrification, supermarket invasion etc. There will also be a FreeZone, where you can bring what you don't need any more and take what you need; an InfoLibrary with literature available about various political subjects; promoting and defending squatting; skill-sharing etc. It would be lovely to hear from you and what you would like to see happening in a space like this. For now and in the future, we are open to your comments and feedback (or complaint!) about what we are doing. The space will officially open on Friday 7th January from 12noon. Come and visit the FreeZone, the library and the info on squatting. From 6pm, we'll have some hot drinks and movies. Everyone welcome!
>
> (Off Market Collective, 2011a)

The call above appeared in early January 2011 in an email circulated on a London-based social centres mailing list. As is often the case with new occupations, the message set out the aims of the collective and listed future familiar uses of the space as an *infoshop*, a *freeshop* and an *open space* to socialise and organise. Occupation-based urban practices such as squatted social centres have long been associated with political processes of resistance to capitalist dynamics and with the constitution of prefigurative alternative urban relations (Vasudevan, 2011; 2015). The spatial appropriation of disused buildings and their transformation into spaces of public and collective use have been studied in the context of the radical political landscape of 'autonomous geographies' in cities across Europe (Montagna, 2006; Ruggiero, 2000; Squatting Europe Kollective, 2013) and in the UK (Hodkinson and Chatterton, 2006). While autonomous urban spaces are at times imagined and represented as 'liberated enclaves surrounded by a hostile capitalist environment' (Stavrides, 2014, p. 547), equating autonomy to distinct spaces 'defined by their exteriority to the rest of the city-society' (ibid.), authors concerned with the transformative power of reclaimed urban spaces as 'urban commons' (Eizenberg, 2012; Newman, 2013) have increasingly paid attention to the politics of interaction of those spaces and practices with the wider city (Stavrides, 2014).

Following the public opening of the OffMarket in a shop front on a busy high street in the North London borough of Hackney, a leaflet was affixed at the entrance and circulated in the local area (Figure 5.1). In contrast to the email, the leaflet was aimed at a wider public beyond the squatting scene and positioned the collective as people who 'live in or near Hackney' and who occupy empty buildings as resources for 'people around us'. If the email presented a set of claims in line with the original intentions to create 'an anarcho-hub in North-East London' (conversation with a member of the OffMarket collective, 14 April 2011), the leaflet's opening question – have you ever thought that there weren't enough public and open places for people to meet, exchange, learn, hang out and get organised? – attempted to address a broader readership. A description of the

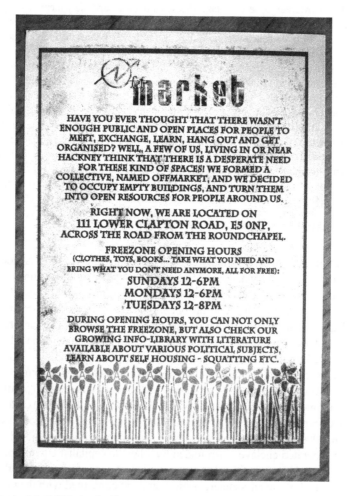

Figure 5.1 OffMarket leaflet, February 2011, front.

Source: Author's personal archive.

FreeZone (also known as freeshop) framed information about the opening hours, drawing attention to an open and regular activity.

With the occupation of a second OffMarket shop, the freeshop became a significant site of engagement with the wider local community, and after its eviction in July 2011 it was described by the core collective as 'hugely successful', partly also because it 'ended up being run mostly by locals' (OffMarket, 2011c). What is at stake in the above claim is the possibility, despite the relatively short life of the two OffMarket shops, of commoning a squatted space beyond the boundaries of intentional squatting and activist communities. In this chapter I analyse this claim by examining the experience of the OffMarket as an instance of 'actually existing urban commons', characterised by multiple, and at times contradictory, modalities and mechanisms of resource sharing in the city (Eizenberg, 2012). Drawing on participant observation and conversations with participants, volunteers and visitors to the two shop fronts, I will examine instances of emerging commoning practices beyond notions of 'liberated enclaves'. Before engaging with the specificities of my case study, in the following section I will outline a few critical issues in recent urban commons scholarship that may be useful to address the challenges and potentials of commoning practices in London.

Reclaiming spaces for urban commoning

From the seminal work of E.P. Thompson on traditions of rural and urban commoning in eighteenth- and nineteenth-century Britain (1991), commons have been understood both as material resources and as the processes of their collective usage and management, negotiated through customs and 'rights' that may not be directly recognised by written law. Peter Linebaugh has expanded scholarship on such customs by critically rereading the origins and development of Magna Carta and its accompanying Charter of the Forest as testimonies of the struggles for common usage in a trajectory of ever-increasing enclosure (2008). Laying the stress on customs and uses, as opposed to rights granted by the state or by another form of authority, Linebaugh has noted that, on several occasions, commons are instituted and reproduced by the sharing of resources through practices of 'commoning', so that commons, as an action, is 'best understood as a verb rather than as "common pool resources"' (2014, p. 8). The praxis of commoning and the resources shared are thus co-constitutive of the commons is concisely summarised by Gidwani and Baviskar, 'commons need *communities*' (2011, p. 42; see also De Angelis, 2003).

The political question of what kind of communities of commoning can be envisaged under conditions of global urbanisation has given rise over the last decade to an increasing body of work. Writing about the concept of the 'metropolis' as one of the key terms introduced in *Commonwealth*, Hardt and Negri have postulated the potential of the urban to be 'a factory for the production of the common' (2009, p. 250), as the site of the activities of production, encounter and antagonism of the political subject of the 'multitude'. Drawing on Hardt and Negri, Chatterton has argued that the 'productive moment of commoning and the social relations that produce and maintain it, is a vital but under-articulated component in our understanding of spatial justice' (2010, p. 627) and has maintained that the vocabulary and imaginary of the commons can offer both material aspirations and organising tools for urban social justice struggles that are 'subversive and oppositional, but also transformative and prefigurative of possible, as yet unknown, urban worlds' (ibid.). In a similar vein, urban commoning has been argued to be integral to a different, more radical understanding of the notion of urban social justice and its implications for urban planning (Marcuse, 2009). Here, as elsewhere (Brenner, Marcuse and Mayer, 2009), the 'right to the city' (Lefebvre, 1996) is invoked as theoretico-political imaginary for establishing action towards collective control and use of urban spaces and resources.

The 'right to appropriation', in particular, has been revisited to stress the potential role of direct action in producing the possibility for collective and transformative use (Knut, 2009). As political theorist Margit Mayer explains, the 'right to the city' is an oppositional right created through social and political action: it's 'a right that exists only as people appropriate it (and the city)' (2009, p. 367). Breaking away from notions of rights as based on national citizenship, critical Lefebvre scholars have argued for an expanded understanding of the political subject of such re-appropriated rights. Mark Purcell has written of a community of enfranchised 'urban inhabitants', membership of which is 'earned by living out the routines of everyday life in the space of the city' (Purcell, 2002, p. 102): a position that, however, has been criticised for risking reproducing 'a view of civil society as basically homogenous' (Mayer, 2009, p. 369) and for neglecting existing class and power divisions.

Debates around the political subject of the 'right to the city', particularly in the context of 'first world urban activism' (Mayer, 2013), provide further analytical tools for the difficult task of examining existing and potential practices of urban commoning as involving the formation of new urban communities through use. While the appropriation of empty urban spaces for common uses can be seen as a powerful embodiment of the right to the city as an oppositional demand – turning a building 'into an open resource for people around us', as in the OffMarket leaflet – there remain the theoretical and political questions of whose needs and desires are met through such appropriations, whose collective and individual subjects are imagined to partake in their on-going commoning and how practices of resource sharing actually take place in reclaimed spaces. In the terse reflections of a member of another squatted social space in Hackney 'are [squatted social centres] "real" community centres? Are "normal" people not from the "scene" coming to them, getting involved and taking part?'(Lou, 2011) And if 'they' do, what kind of relationships to the space and its resources are established?

Drawing on Jacques Rancière's writings about thresholds as 'artifices of equality' (2010), Stavros Stavrides has proposed to lend renewed attention to what he terms the 'threshold spatiality' of occupations. With the notion of 'thresholds' he designates the ability of reclaimed spaces to host and express 'practices of commoning that are not contained in secluded worlds shared by secluded communities of commoners. Thresholds explicitly symbolize the potentiality of sharing by establishing intermediary areas of crossing, by opening the inside to the outside' (2014, p. 547). Echoing what Jenny Pickerill and Paul Chatterton have called the 'power of interaction with society' of autonomous practices (2006, p. 741), Stavrides argues that in order for commoning to remain 'a force that produces forms of cooperation-through-sharing, it has to be a process which overspills the boundaries of any established community, even if this community aspires to be an egalitarian and anti-authoritarian one' (2014, p. 548). The idea of commoning thresholds raises two interrelated issues: on the one hand, a question of communication and mediation of the openness of reclaimed spaces, beyond 'secluded communities of commoners'; on the other, that of the relationship between the idea and practice of sharing 'wasted urban resources' and the social and economic positions of the users and volunteers in the space, with differing degrees of privilege and dispossession, particularly at times of reduced public spending and increasing inner-city poverty.

Focusing on the experience of the OffMarket, I will begin to address the first issue by analysing the ways in which the space presented itself through its 'approachable aesthetics' and its performative 'staging' of openness. While scholars agree on the important distinctions between notions of 'the commons' and of 'public', the latter understood a juridical category pertaining to the state and the law, and, defined in opposition to the 'private' (Hardt and Negri, 2009; Gidwani and Baviskar, 2011), the study of threshold spatialities of commoning could benefit from a critical discussion of intentional 'public' openness as socially and spatially emergent through processes of mediation and use (Mahony, Newman and Barnett, 2010, see also Iveson, 2007). Specifically, my participation in the freeshop as a volunteer will form the basis of my examination of the 'free' exchange of objects as a strategy of openness to a wider 'public' in the commoning of the space.

In the second part of the chapter I will focus on significant moments of the lived experience of commoning in the space, in order to think further about the multiple thresholds of the space, embodied in everyday encounters and negotiation of uses with the invoked but elusive subject of the 'local inhabitants'. Through a situated reflection on accounts of the mundane and, at times, emotional exchanges through which 'a sense of the common is produced' (Dawney, 2013: p. 149), I aim to examine the potential and difficulties of commoning in an inner-city London high street. Beyond the seizure of common spaces there is widespread agreement that commons are made and remade through different kinds of work (Eizenberg, 2012; Gidwani and Baviskar, 2011), which need to be addressed in themselves and in relation to wider urban dynamics. In the final section I will therefore examine the diverse economies that enabled the sustenance of the freeshop and the practices of commoning that constituted and expanded networks of participation beyond the physical space. I will conclude by raising three critical reflections on the study of the OffMarket experience and by discussing wider implications for a more expanded understanding of the transformative mechanisms of urban commoning beyond established communities.

'Approachable' social centres

My first visit to the OffMarket, prompted by the email of the opening quote, challenged my expectations. Despite its location on a local high street and an A-board sign with the words 'We are open COME IN!' I missed it, as the red shop front visually harmonised with surrounding independent barbershops, charity shops and cafeterias (diary entry, 31 January 2011) (Figure 5.2). My impression of the space's 'mimetic' appearance was shared by other visitors (conversation, 14 February 2011), including a local long-term resident who mentioned how he had seen the shop and decided to walk in with his six-year-old daughter, something that he admitted wouldn't have happened if it had looked like a 'usual squat' (conversation, 15 March 2011). In this sense, the OffMarket could be associated to a number of 'approachable' community-oriented political spaces, occupied in East and North London between 2009 and 2011, in contrast to 'the grimy punk attitude that pervades some other activist spaces' (Maxigas, 2009). The approachable aesthetic was not accidental, and derived partly from the desire to open the space to passers-by and local residents and partly from the wish to align it more to the tradition of radical infoshops (Dodge, 1998) rather than to 'party' or 'crash' spaces – an important distinction made by some self-declared social centres (Lou, 2011).

After the first OffMarket was evicted in late February 2011, another shop was reclaimed a few metres from the first one and rearranged with a similar appearance. The new space was situated between a vacant shop, a hardware shop, a charity shop and a launderette, near a large 24/7 off-licence shop and overlooking a zebra

Figure 5.2 Front of first OffMarket shop, February 2011.

Photo by author.

crossing. The interior of the front space was organised around a sitting area with a sofa, two armchairs, two coffee tables and a tea corner above a small fridge, behind which was a foldable table covered with stacks of leaflets and flyers. On the left wall there was a plain-words definition of anarchism as non-hierarchical self-organisation, while on the right-hand side was affixed a large sheet of paper with the question 'What would you like to see/do in this space?' followed by a list of suggestions and proposals, with email addresses of interested people (diary entry, 14 February 2011). Leaning against the shop front window there was a board with a text that explained that the space was squatted, that popular media representations of squatters are 'sensational crap that papers publish to sell more and divide us', and that concluded with the sentence 'come in, we don't bite and we like sitting around with a cuppa!!' (Figure 5.3).

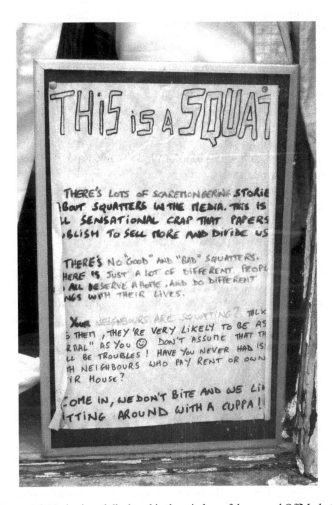

Figure 5.3 Notice board displayed in the window of the second OffMarket shop, April 2011.
Photo by author.

A large section of the shop was occupied by the 'freeshop', an area where clothes, tools, books, shoes and other objects could be donated, taken for free or swapped. The freeshop was located on the right-hand side of the front room, but during open days it would expand to occupy most of the space both inside and outside the shop. The objects on display varied from day to day, and included used children's clothes, books, CDs, bedding, fabric cuttings and shoes. On open days the freeshop spilled out onto the pavement, often with the railing of children clothes, a bench and a supermarket trolley containing second-hand VHS cassettes. Above the railings on the left wall, a sign acted as textual framing: 'This is a free shop. Take what you need and bring what you don't need any more!' Even with a high turnover of objects, the freeshop was constantly replenished, which evidenced the support of local residents, some of whom chose to donate to the freeshop rather than the adjacent charity shops (diary entry, 7 June 2011).

Despite the informal, colourful and 'messy' visual experience, the use of the space was subjected to a series of rules, some of which were written on posters affixed to the walls, such as 'no cameras please' and 'no smoking', while others were included in a rather lengthy 'safer spaces policy' also displayed in the main room, which had been devised by the core collective before opening the space and which was grounded on the principle of 'making sure everyone felt welcomed' (OffMarket, 2011c). Other rules were agreed through collective discussions after specific incidents, such as for instance the decision not to allow unaccompanied young children in the space. While this was an indication of the extent to which the space was perceived as safe and friendly by the parents, it put volunteers in a difficult position, both legally and in terms of the safety of the children in case of unexpected violence from the owners or a raid by the police.

'Where's the trick?'

The open door afternoons at the freeshop were an important time for encountering and establishing relations of commoning with visitors and passers-by. As explained by a member of the collective, a freeshop in a social centre is 'an easy solution' to the question of how to engage with a wider public 'because anybody can do with free stuff' (conversation 14 April 2012). But while freeshops are part of the expected repertoire of occupied spaces, the 'approachable' appearance of the OffMarket and the fact that it was an actual shop front required forms of performative explanations. A common reaction of newcomers was to ask who was 'behind' the space and who managed it. During one of my initial visits to the space, an elderly man with a foreign accent entered the shop with two large carrier bags and asked 'Who is in charge here?' to which another volunteer replied, smiling, 'Nobody!' Amused and apparently satisfied with the answer, the man laughed and left a bag full of men's clothes and porcelain knick-knacks (diary entry, 14 February 2011). This vignette offers an example of everyday questioning of the freeshop, and the semi-serious answer, which complemented the explanatory writings on the walls, was an important self-identifying utterance that pointed to the prefigurative practices of a squatted 'free' space.

These performative explanations were considered very important to the framing of the space. All volunteers were given a booklet titled *A little guide to OffMarket* setting out basic rules and tasks. Beyond everyday maintenance and making 'the place look nice' by drawing open the curtains, opening the entrance door and placing on the outside pavement boards and signs to welcome visitors and inform them of weekly activities (OffMarket, 2011b), interaction with visitors was explicitly encouraged, as volunteers were expected to 'make people welcome when they come in' and 'check if they know what the place is about and if not explain them quickly'. On busy days, passers-by could be seen walking in from the street, stopping in the middle of the space, looking around disoriented, perhaps realising that they had meant to enter the adjacent charity shop; the mimetic appearance of the shop front was, to a certain degree, misleading. In some circumstances the misunderstanding continued until they approached one of the volunteers with an object and asked 'how much?' The answer 'it's free' would always create a moment of disbelief and suspicion. As observed by a volunteer at another freeshop, 'if you work in a freeshop, you have to explain a lot about anarchy and freeconomy' (Maxigas, 2009).

In the words of a freeshop volunteer, the main reply was 'where's the trick?' (conversation, 14 April 2012). At this point, volunteers would usually explain that there wasn't any 'trick': the shop had been vacant for many years and was wasted, so it had been reclaimed and was now free for social uses; the clothes, the books and other goods on display were similarly free because someone, somewhere, did not need them any more. For some freeshop volunteers this particular speech-act constituted the 'magical moment' (conversation, 27 June 2011) when the commoning logic of the space was understood by the newcomer, enabling the transition from suspicion and surprise to complicity and solidarity. In different variations, the performative iteration of the question and its answer could be seen as a form of 'ethical spectacle' in which 'the texture of the experience [. . .] the emotions generated and felt – and the autonomy of the action matter[ed] as much as its political outcomes' (Routledge, 2010, p. 208, following Duncombe, 2007). The performance of 'freeness' and 'openness' relied on the initial misreading of the shop as a 'regular' shop, and on the suspicion and disbelief of the interlocutor, so as to score a political point about the wastefulness of capitalism and its logic of enforced scarcity.

Conversations and shared work between members of the collective, volunteers and marginalised local residents became more frequent as the freeshop gained in popularity as a resource and as a free space for socialising. Besides squatters and activists, a large proportion of everyday users and visitors to the OffMarket freeshop were long-term unemployed, working-poor, undocumented migrants, pensioners and vulnerable individuals with physical and mental disabilities. Some received different forms of welfare support, about which they expressed anxiety, lived in overcrowded situations or were limited in their movements around the neighbourhood by ASBOs (Anti-Social Behaviour Orders) and dispersal zones. For some local inhabitants, the shop had become a first point of contact, either when they did not know where to go or – as was more often the case – when they had been turned away from or disappointed by social services or charities (diary entry, 5 July 2011). The local popularity of the freeshop revealed the local high levels of social and economic deprivation, which became particularly visible city- and world-wide a few weeks after the eviction of the second OffMarket with the street looting and rioting that took place a few hundred metres away during the London Riots in August 2011.

The question of the space's turning into a stop-gap solution was particularly visible in encounters with migrants, low-income pensioners and elderly people, who at times entered the shop under the mistaken impression that it was a regular charity shop. On one such occasion, a frail elderly woman reacted to the explana- tion that all objects were free by holding my hand, with watery eyes, thanking me profusely and saying that the space would bring a lot of luck to the person behind it. Her hand felt very thin, and I felt initially touched and shaken, and sub- sequently uncomfortable and angry about having 'tricked' her into tears of grate- fulness (diary entry, 14 June 2011). When describing this incident to a member of the collective, we agreed that the 'mimetic' appearance of the shop front was successful in attracting people who might have not entered a squatted social centre otherwise. In the case of some elderly and vulnerable visitors, however, it was clear that the non-monetary exchange did not by itself institute radically different and transformative relations, and for those individuals the shop was 'just another charitable place where for some reason things are free [and most would] just take without thinking too much' (conversation, 14 April 2012).

The format of occupied 'social centres' is often presented as a clear instance of commoners' struggle because of the ways in which people 'get together and seize common spaces and turn them into projects of welfare from below' (De Angelis, 2007, p. 4). Yet the provision of welfare 'from below' can lead to situations that are 'hard to deal with' by untrained volunteers, as was realised by participants in the autonomous social centre The Cowley Club in Brighton, who discovered that local social services were encouraging vulnerable people to go to the club, which was providing cheap meals, advice, language classes and space for socialising, raising the question of whether open social centres are 'more a stop gap in social services than a radical solution to society's problems' (The Cowley Club, 2007, p. 23). In the case of the OffMarket, free clothes and a warm, dry place to socialise with free tea and biscuits were less of a countercultural alternative to everyday capitalist consumption and more of a direct answer to a material necessity, at the margins of but not inherently antagonistic to or critical of capitalist urban dynamics.

Moreover, the mechanism of alleged 'freeness' gave visibility to unspoken assumptions about social and economic differences among participants in the free- shop. While in the intentions and desires of volunteers and users of the space there was a commitment to forms of mutual aid, solidarity and horizontality, relatively more privileged individuals tended to donate but hesitated to take objects (conversa- tion, 17 June 2011; conversation 12 March 2012). In this, pre-existing cultural and social assumptions about free donations as 'charity' seemed to shape both sides of the act, fixating the subjectivities in the freeshop interactions into learned patterns of givers or receivers. My own initial tendency to take on the role of giver towards 'needier' users and volunteers was indirectly challenged by a conversation with another one of the main organisers of the freeshop, a woman in her early forties, who lived locally in sheltered housing. During our shifts she often took garments and spare pieces of fabric to tailor models that she would sell to sustain herself and her family, proudly saying that she had always worked and that if she needed welfare support it was not for lack of trying. Noticing my reluctance, she encouraged me to take anything I wanted, just like she did, because it was 'free for everyone' (diary entry, 31 May 2011). In this as in other instances, working side by side with vulnerable local inhabitants challenged my own reason and position within the performance of freeness, raising questions about unspoken 'charitable' assumptions in the uncomfortable and at times emotional engagement with the 'outside'.

Commoning off the market

Beyond the encounters occurring through the activities of the freeshop, the commoning of the OffMarket included a wide array of resources and participants in their use. As indicated by the name, a desire for autonomy from capitalist market relations was implicit in the aims and forms of sustenance of the space, which presented itself as a prefigurative example of re-appropriating and collectivising 'wasted' resources, from vacant spaces, to clothes, to food and objects of everyday use. To encourage the sharing of knowledge and skills about squatting there was a map of empty houses and a system of codes for identifying useful information and issues. Social evenings and regular open cafes offered vegan meals and snacks prepared from 'skipped' food, and another map of London was on display with a list of accessible sites, with the best days and times. A similar logic of reusing 'waste' was central to the weekly bicycle repair workshops, where spare parts were exchanged and users could swap skills and use donated tools collectively.

The commoning activities in the space were supported by mixed diverse economies of donations, non-monetary exchanges, voluntary labour and in-kind support, as well as peripheral market exchanges. To ensure non-monetary exchanges, the 'donation box' (which was mostly used for paying for the collective mobile phone) was usually hidden on open days, to avoid visitors and users offering money in exchange for the objects they were taking or for their use of tools, which would have undermined ideas of freeconomy and mutual aid (OffMarket, 2011c). In the specific case of the freeshop, however, participants were encouraged to wash donated clothes at home and return them to the space, thus relying on capitalist forms of laundry processes. If anyone insisted on making a monetary donation, volunteers suggested offering supplies for the 'tea corner', such as dry biscuits, sugar, milk or tea bags, which meant engaging with monetary market exchanges with local corner shops. The tea area was to be replenished by volunteers or by members of the core collective, which in practice assumed a shared ability to contribute one or two pounds each shift to the running of the space. During one of my first shifts in the space, however, I overheard a regular volunteer, an unemployed woman in her forties, mentioning how she couldn't contribute to the pint of milk because 'at the end of the week it all add[ed] up' (diary entry, 31 May 2011). Just as with the freeshop, the tea corner was never depleted, which seemed to indicate an unspoken understanding, at least among some volunteers, that those who could should contribute more, a contribution that was rarely 'off the market'.

Volunteers and users were also involved in the organisation and programming of events and activities. In terms of its self-organisation, the space worked on a permeable three-tier system. There was a core collective of people who were most intensively involved; a crew of regular volunteers committed to regular activities in the space; and, finally, individuals or groups who used the space for specific one-off events, such as an external meeting or film screenings. The tasks undertaken by members of the core collective included writing and distributing the weekly email newsletter, checking and responding to the space's emails, meeting and training new volunteers, ensuring that the space was clean and safe, and managing the occupation rota which sometimes also involved individuals from the volunteer crew. Decisions regarding the programme and the running of the space were taken during the weekly evening open meetings and involved the core collective, volunteers and anyone who had come to propose new activities. As declared in another leaflet: 'We try and work in a collective and horizontal way, where the people affected by the decisions taken are actually the ones taking them (i.e. not like a parliament!) and we do our best to oppose all discriminations and hierarchies that are set up between us!' Decisions about activities in the space were made by consensus among the people present and one member of the core collective on rotation would take responsibility to check space availability, liaise for access and promote the activity through the weekly newsletter.

The success of the 'freeshop' in turning the shop into a social space and a free resource was at times seen by members of other collectives within the space as detracting from the overall aims of the squat to promote a wider range of political activities and to involve the 'freeshop' visitors in other uses of the space beyond 'hanging out' and exchanging objects. Besides the decision to limit the physical space dedicated to objects within the shop, as well as the opening hours, this reflection also led to a conscious effort, whenever possible, to accompany explanations of free use with an active encouragement to participate in existing activities, or to propose new common uses by attending the weekly meetings. For instance, it became apparent that the freeshop was very popular among parents of young children, as evidenced by the rapidity with which the shelves with children's and babies' clothes and toys needed refilling, and by the regular requests for more (diary entry, 31 May 2011). After sharing this observation with the freeshop volunteers and with other young parents, a local mother in her thirties started an email list on a piece of paper hung by the children's section, with the aim of organising collective free childcare. The list became popular as volunteers actively promoted it to visitors with children or browsing for children's clothing and toys, leading to the organisation of a first public meeting, which unfortunately was cancelled when the second OffMarket was evicted.

Beyond activities of commoning in the shop itself, the OffMarket attracted wide local support, as evidenced by many mundane yet important acts of solidarity by traders and residents in the street and surrounding areas. A local Indian food takeaway, for instance, introduced unofficial discounts for older people after an unemployed elderly volunteer explained to them why he was spending his days in the space. A nearby RSPCA charity shop donated, unsolicited, a clothes rack that it was not using. When the squatted shop was raided by the London Metropolitan police on the eve of the royal wedding and two volunteers were detained, local traders and residents visited the space to help sweep away the shattered glass of the door and expressed sadness and anger at the violence of the operation (diary entry, 27 June 2011).[1] With time, volunteers and members of space had began to engage more actively in activities beyond the shop, such as a running a 'freeshop' stall in the grounds of a local church hall during a neighbourhood event. This in turn had attracted an invitation to hold a regular 'freeshop' stall for free at the weekly resident-led food and second-hand market 'Homemade Hackney' in the grounds of another local church. The stall became popular and often attracted donations from other stall holders (diary entry, 2 July 2011), showing how the prefigurative practices of the OffMarket were beginning to engage with and establish relations of non-monetised commoning beyond the shop front. After the eviction in mid-July 2011 a local resident and regular volunteer of the freeshop carried on running the 'OffMarket stall', symbolically and physically continuing to reclaim a space for a diverse form of sharing. Despite local support, the stall stopped in the middle of August, partly because of the loss of a space for collecting and storing objects, partly because of more urgent social and political activities surrounding the rioting in Hackney that same summer.

Conclusion: thresholds of commoning

> Commoning is exclusive inasmuch as it requires participation. It must be entered into.
>
> (Linebaugh, 2014)

> But do we really need to 'offer' something to people? to provide services? to attract them thinking that we could convert/enlighten/radicalise them? Are people around us not intelligent enough to just take and find what they want, where and when they want?
>
> (Lou, 2011)[2]

Temporary and small scale-acts of commoning have been argued to enable both a vantage point from which to see more clearly capitalist modes of relations and the possibility to experience forms of everyday commoning that make 'an *outside* dimension to the value practices of capitalism *visible*, by virtue of our being there and declaring our presence as *other*' (De Angelis, 2007, p. 24, original emphasis). Despite their relative temporariness, the two OffMarket shops exposed multiple forms of 'waste' in the city, from spaces, to food, to objects, and the possibility to reclaim and re-appropriate them through commoning. The shop embodied them and was also a space of common learning where knowledge and skills about skipping and squatting, as well as about recycling and sharing objects of everyday use, could be shared and improved. The space itself was set up as an open resource that actively sought to expand familiar activist activities with events and proposals from visitors and participants with little prior knowledge of political squats.

The chapter has examined the ways in which the OffMarket's freeshop functioned as a site of encounter and engagement with urban inhabitants beyond the anarchist and squatting 'scene' and has attended to the practices of framing the space as 'open' and of translating the 'magic' of direct action into a collective reclaiming of the means of social reproduction in urban settings. In 'spilling over the boundaries' (Stavrides, 2014) of established activist communities to recognize and include other spaces and practices of 'off the market' relations, the experience of the OffMarket explored an expanded understanding of the commoning potential of the neighbourhood. While there remained substantial differences in involvement and practical knowledge between the core collective and the groups of volunteers for each regular activity, on the one hand, and visitors, local traders and occasional users, on the other, the multiplicity of activities and more everyday relations established through the use of the space offered a significant short-lived enactment of several 'rights to appropriate' the city through commoning. In this conclusion I draw on the experience of the OffMarket to raise three critical reflections concerning power, space and the making of the commons in contemporary urban conditions.

The first reflection concerns forms of urban commoning at times of increasing personal and collective impoverishment as well as of enclosure and privatisation of social services and community spaces. One of the core characteristics of 'actually existing' urban commons is that that it should fulfil social needs – such as open space, recreational and social spaces – in a non-commodified manner (Eizenberg, 2012, p. 766). The aspiration to enact non-commodified ways of sharing resources, however, is inevitably complicated by the multiple 'actually existing' social and economic relations that co-exist to enable a project of common sharing, and by their relationship to neoliberal capitalist urban dynamics, particularly at a time of 'urban austerity'. Critically, the freeshop as a device to redistribute wasted objects and clothing could be seen as a residual and parasitic practice within capitalist market relations, reproducing a scenario in which 'commons (and the communities that sustain them) are relay points in the social life of commodities' (Gidwani and Baviskar, 2011, p. 43), subsidising and supplementing capital accumulation, rather than radically challenging it.[3]

Similarly, as observed by Caffentzis, the provision of 'welfare from below' (De Angelis, 2007) through voluntary labour and self-organisation of the commons should not be considered as inherently anti-capitalist, and could actually be seen as unwillingly contributing to strategies of austerity-led 'neoliberalism plan b' (Caffentzis, 2010) in a specific British context of 'austerity localism' (Featherstone et al., 2012). These contradictions do not concern simply the more abstract level of economic and social autonomy, but also the more embodied and at times emotional level of one-to-one interactions, as evidenced in the analysis of the 'charitable' encounters with the elderly woman and in the unspoken assumptions of some of the donors and volunteers. The tension between individual and collective ethical and political intentions and their role as providers of residual welfare may be resolved only by a more critical and situated political understanding of the difficulties of shifting from modes of 'provision' to more fundamentally transformative moments of commoning, where activities of 'welfare from below' are collectively undertaken and shared. On this point, the attempt to co-organise collective free childcare in the space, albeit unsuccessful, revealed a shift in low-income parents' relation to the 'open resources' of the OffMarket: from a site of socialisation and objects-swap into a place capable of accommodating the recognition of common needs and desires that had not been anticipated by the core collective, and the capacity to act on it collectively.

Following this, a second reflection on the potential and limitations of reclaimed spaces concerns the question, raised in the introduction, of the collective subject of urban commoning and of the 'exclusivity' of participating, to paraphrase Linebaugh's quote. In the case of the OffMarket, the inevitable differences between the core collective, the more relatively privileged volunteers and donors, and the most vulnerable users and visitors to the space presented significant challenges to the politics of the space, its 'freeness' and openness. The 'trick' of direct appropriation and commoning of wasted urban resources highlighted frictions between the desire to establish an autonomous space of reciprocity and commons, and the multiple deprivations of vulnerable groups in an inner-city borough. Rather than seeing these frictions as instances of failed commoning, they may be considered as an inevitable and welcome starting point for practices that hope to create linkages between often relatively privileged activists and the more vulnerable and marginalised urban inhabitants through 'a shared awareness of the importance of critical reflexivity about class and privilege' (Mayer, 2013, p. 17).

Even with the development of a shared critical reflexivity, however, there remain the tensions identified in the second quote at the beginning of this section between a politics of urban appropriation that wants to attract people in order to 'convert/enlighten/radicalise them' and the desire for a more equal and reciprocal relationship with the 'outside', based on an understanding of other urban inhabitants as capable to take and make their own commons 'where and when they want'. In response to the quote, perhaps the question is less about the collective 'intelligence' of non-activist local inhabitants, in its original sense of being able to grasp the legibility of the situation they live in and their needs, and more about the availability of time and resources, including skills and knowledges, necessary for beginning to conceive of collective forms of resource sharing, let alone to implement direct appropriation.

My final reflection concerns the ways in which emerging geographies of urban commoning can be conceptualised and researched. If thresholds are 'powerful tools in the construction of institutions of expanding commoning' (Stavrides, 2014, p. 547), there is a need for a greater and more nuanced understanding of their spatialities, their emotional and material embodiment and the complex and dynamic negotiations over framings and meanings that are produced through collective open use. Methodologically, such a project would require an expanded understanding of the site of such autonomous geographies, both spatially and temporarily, moving from the study of bounded place occupations to include multiple sites and processes of self-organisation, solidarity and resource sharing that may be established across profit and not-for-profit capitalist activities in the city. It would also require a far greater attention to sustained relationships as well as to occasional mundane commoning interactions beyond the 'subject' of established activist collectives, to include encounters with urban inhabitants who may only occasionally brush against commoning practices. This study is an attempt to explore this approach by explicitly acknowledging the negotiations, misunderstandings and discomforts of new emerging social and political configurations, to catch glimpses of multiple shared practices of urban commoning in which the separation between 'us' and 'them' is challenged and struggled over.

Notes

1 The raids affected several squatted social centres across London and were condemned by the urban social movement community as being motivated by a desire to intimidate spaces of political dissent (FIT Watch, 2011).
2 Blogpost by an activist reflecting on her experience of another squatted social centre in North-East London.
3 The dialectic of waste in relation to modes of capitalist accumulation and the possibility of commoning (see, for instance, Gidwani, 2013) would warrant a more sophisticated analysis than is possible in this text.

References

Brenner, N., Marcuse, P. and Mayer, M. (2009). Cities for people, not for profit: introduction, *City: Analysis of Urban Trends, Culture, Theory, Policy, Action* 13(2/3), pp. 176–184.

Caffentzis, G. (2010). The future of 'the commons': neoliberalism's 'Plan B' or the original disaccumulation of capital?" *New Formations*, 69, pp. 23–41.

Chatterton, P. (2010) Seeking the urban common: furthering the debate on spatial justice, *City: Analysis of Urban Trends, Culture, Theory, Policy, Action*, 14(6), pp. 625–628.

Cowley Club, The (2007), in Social centre Network (ed.) *What's this place? Stories from radical social centres in the UK and Ireland.* Leeds: Autonomous Geographies, http://socialcentrestories.wordpress.com/ [accessed 7 July 2011].

Dawney, L. (2013). Making common worlds: an ethos for participation. In Noorani, T., Blencowe, C. and Brigstocke, J. (eds) *Problems of participation: reflections on authority, democracy, and the struggle for common life.* Lewes: ARN Press, pp. 147–151.

De Angelis, M. (2003). Reflections on alternatives, commons and communities or building a new world from the bottom up. *The Commoner* 6, pp. 1–14.

De Angelis, M. (2007). *The beginning of history: value struggles and global capital.* London: Pluto Press.

Dodge, C. (1998). Street libraries: infoshops and alternative reading rooms. web.archive.org/web/20091027155038/http://www.geocities.com/SoHo/Cafe/7423/infoshop.html [accessed 1 Oct 2015].

Duncombe, S. (2007). *Dream: reimagining progressive politics in an age of fantasy.* London: The New Press.

Eizenberg, E. (2012). Actually existing commons: three moments of space of community gardens in New York City. *Antipode* 44(3), pp. 764–782

Featherstone, D., Ince, A., Mackinnon, D., Strauss, K. and Cumbers, A. (2012) Progressive localism and the construction of political alternatives. *Transactions of the Institute of British Geographers* 37, pp. 177–182.

FIT Watch (2011). Police raids on squats and social projects (29 April), www.fitwatch.org. uk/2011/04/29/police-raids-on-squats-and-social-projects/ [accessed 15 June 2011].

Gidwani, V. (2013). Six theses on waste, value, and commons. *Social & Cultural Geography* 14(7), pp. 773–783.

Gidwani, V. and Baviskar, A. (2011). Urban commons. *Economic & Political Weekly* 46(50), pp. 42–43.

Hardt, M. and Negri, A. (2009). *Commonwealth*. Cambridge, MA: Harvard University Press.

Hodkinson, S. and Chatterton, P. (2006). Autonomy in the city? Reflections on the social centres movement in the UK. *City: Analysis of urban Trends, Culture, Theory, Policy, Action* 10, pp. 305–315.

Iveson, K. (2007). *Publics and the city*. Oxford: Blackwell.

Knut (2009). 'Right to the city' as a response to the crisis: 'convergence' or divergence of the urban social movements? www.reclaiming-spaces.org [accessed 13 April 2011].

Lefebvre, H. (1996) *Writings on cities*. Oxford: Blackwell Publishers.

Linebaugh, P. (2008) *The Magna Carta manifesto: liberties and commons for all*. London: Verso.

Linebaugh, P. (2014). *Stop, thief! The commons, enclosures, and resistance*. Oakland, CA: PM Press.

Lou (2011). A personal account on 195 Mare St squat, Indymedia London, http://london. indymedia.org/articles/7726 [accessed 22 April 2011].

Mahony, N., Newman, J., and Barnett, C. (2010). *Rethinking the public: innovations in research, theory and politics*. Bristol: Policy.

Marcuse, P. (2009) From justice planning to commons planning. In Marcuse, P., Connolly, J., Novy, J., Olivo, I., Potter, C. and Steil, J. (eds) *Searching for the just city: debates in urban theory and practice*. London: Routledge, pp. 91–102.

Maxigas (2009). Another buy nothing day at the Freeshop. Indymedia London, http://lon-don.indymedia.org/articles/3236 [accessed 12 March 2011].

Mayer, M. (2009). The 'Right to the City' in the context of shifting mottos of urban social movements. *City: Analysis of Urban Trends, Culture, Theory, Policy, Action* 13, pp. 362–374.

Mayer, M. (2013). First world urban activism: beyond austerity urbanism and creative city politics. *City: Analysis of Urban Trends, Culture, Theory, Policy, Action* 17(1), pp. 5–19.

Montagna, N. (2006). The de-commodification of urban space and the occupied social centres in Italy. *City: Analysis of Urban Trends, Culture, Theory, Policy, Action*, 10(3), pp. 295–304.

Newman, A. (2013). Gatekeepers of the urban commons? Vigilant citizenship and neoliberal space in multiethnic Paris. *Antipode* 45(4), pp. 947–964.

OffMarket Collective, (2011a) #OffMarket# new squatted space opening Friday 7th, Rampartdiscuss e-mailing list, 3rd January 2011.

OffMarket (2011b). *A little guide to OffMarket*.

OffMarket, (2011c). OffMarket: evicted space, project still alive!, Indymedia London, 12 July 2011 [accessed 10 September 2011].

Pickerill, J. and Chatterton, P. (2006). Notes towards autonomous geographies: creation, resistance and self-management as survival tactics. *Progress in Human Geography* 30, pp. 730–746.

Purcell, M. (2002). Excavating Lefebvre: the right to the city and its urban politics of the inhabitant. *GeoJournal* 58, pp. 99–108.

Rancière, J. (2010). *Dissensus. On politics and aesthetics*. London: Continuum.

Routledge, P. (2010). Dreaming the real: a politics of ethical spectacle. In Birch, K. and Mykhnenko, V. (eds) *The rise and fall of neoliberalism: the collapse of an economic order?* London: Zed Books, pp. 206–221.

Ruggiero, V. (2000). New social movements and the 'centri sociali' in Milan. *The Sociological Review* 48(2), pp. 167–185.

Squatting Europe Kollective (ed.) (2013). *Squatting in Europe: radical spaces, urban struggles*. New York: Minor Compositions.

Stavrides, S. (2014). Emerging common spaces as a challenge to the city of crisis. *City: Analysis of* urban trends, culture, theory, policy, action 18 (4–5), pp. 546–550.

Thompson, E.P. (1991 [1967]) *Customs in common: studies in traditional popular culture*. New York: New Press.

Vasudevan, A. (2011). Dramaturgies of dissent: the spatial politics of squatting in Berlin, 1968-. *Social & Cultural Geography* 12, pp. 283–303.

Vasudevan, A. (2015). The autonomous city. Towards a critical geography of occupation. *Progress in Human Geography* 39, pp. 316–337.

Part III
An expanded commons

6 Expanding the subject of planning

Enacting the relational complexities of more-than-human urban common(er)s

Jonathan Metzger

Introduction: what should have a right to the city?

In a text outlining a proposition for urban 'commons planning', the respected critical urban theorist Peter Marcuse (Marcuse, 2009a) opens with an epigraph of the first stanza of an English anti-enclosure folk poem from the seventeenth century:

> The law locks up the man or woman,
> Who steals a goose from off the common,
> But leaves the greater villain loose
> Who steals the common from the goose.

In the continuation of Marcuse's text it becomes obvious that, for him, the goose merely plays the role of a passive prop in his continued reflections on the urban commons and their distribution, which instead becomes implicitly framed as an exclusively and exclusionary human problem. It appears as if, to Marcuse, the 'problem of the commons', of how to best care for the commons in general – and urban commons specifically – was self-evidently to be understood as a question concerning how to divide up the benefits of the commons between various *human* groups, while the aforementioned goose is unceremoniously ushered to exit the discussion stage left.

In this chapter I aim at destabilizing this generally taken-for-granted basic assumption in discussions on urban commons. I ask myself: what happens if we read the above-cited stanza differently, and perhaps somewhat more literally, to also consider that the goose herself might actually hold a legitimate claim to the commons, including any specifically urban such? I further propose that such a reconceptualization of the commons might be completely necessary in the light of the current challenges facing us humans as a species, presently becoming identified, through its causes, as the *Anthropocene* (Zalasiewicz et al., 2010), or – more dramatically – through its consequences, as the *Sixth Extinction* (Barnosky et al., 2011). What both these terms highlight is how we humans, collectively as a species (albeit, some of us drastically more than others), at present appear to be busy not only undermining our own planetary preconditions of existence but, further, continuously – to greater or lesser extents – perpetrating the 'greater villainy' of stealing the commons on a planetary scale from the goose and myriads of other beings and existences, and thereby also generating a perverted feedback loop through which we even more rapidly accelerate the pace of our own probable demise.[1]

The line of reasoning presented in this chapter is based on a recognition that it is now completely necessary for humans, particularly global Northern, well-to-do urbanite members of our species, to begin to grapple with the ethical and practical implications of this insight so as to try to find ways to develop a new, less (self-) destructive sensibility towards the world, more befitting our current predicament. By bringing this somewhat dauntingly overwhelming ambition home to one of my specific areas of interest, questions of urbanity, I hope to be able to make it somewhat more tangible and concrete. In the context of the above-presented overarching ambition, 'the urban' also appears to present a particularly interesting area of investigation almost of a limit-case character, since the City (capital 'C') historically by and large has been imagined and conceptualized as the exclusively human domain *par excellence*, notwithstanding all the motley and variegated myriad creatures that have inhabited any and every really existing city (minor 'c') throughout the world (Metzger, 2015).

The lack of respect for, or even recognition of, the other-than-human denizens of cities is an on-going affair which is, unfortunately, also reproduced in and through the mainstream of contemporary urban planning theory (although there exist notable exceptions to this, see e.g. Wolch et al, 1995; Wolch, 2002; Hinchliffe & Whatmore, 2006 and Holmberg, 2013). Critical planning theory, such as that of Marcuse, has provided seminal contributions towards prompting planners to more explicitly ask the question *cui bono?* – who benefits? – in relation to their ongoing practices, and to stop hiding behind a hollow legitimization of claiming to represent some form of phantom 'public interest'. Certainly the slogan that Marcuse unreservedly propagates, 'Cities for people, not for profit' (Marcuse 2009b: 195), chimes well in devout humanist ears. But – with the above line of reasoning in mind – is it really so self-evident that cities should be only for 'people'? In my view, by focusing so exclusively on humans, the current mainstream of critical planning theory reproduces certain taken-for-granted truths about the world as commonsensical in a highly problematic and even dangerous way. Therefore, I think this is the time to question whether it is really reasonable to sustain the misapprehension that urban development is an activity which affects only humans, and that humans should be the sole beneficiaries (and victims) that deserve consideration in the development of urban settlements.

To some degree this then amounts to a reiteration of the question Isabelle Stengers has asked herself in a different context: 'how to design the political scene in a way that actively protects it from the fiction that "humans of good will decide in the name of the general interest"? How to turn the virus or the river into a cause for thinking? But also how to design it in such a way that collective thinking has to proceed "in the presence of" those who would otherwise be likely to be disqualified as having idiotically nothing to propose' (Stengers, 2005: 1002). As argued by Castree (2003: 207), such an approach would first demand that we abandon the traditional idea that political rights, entitlements and deserts apply to only people; second, confront the very real problem of defining political subjects in a world where the boundaries between humans and non-humans are hard to discern; and third, expand political reasoning to include non-humans, yet without resorting to the idea that the latter exist 'in themselves'.

In this text I will try to make a small contribution towards working these insights into contemporary planning theory and methodology, particularly in relation to the idea of 'urban commons' and, relatedly, '(urban) commons planning'. In line with the above, my main errand in this text is to try to figure out what could potentially become of ('critical') planning theory and methodology if we attempt to reconfigure it by taking our cue from *more-than-human* thinking, so as to radically expand the scope of beings that 'count' (cf. Whatmore, 2002) and which we also may be accountable before in our actions and their consequences. What new trajectories towards the future might this open up for the highly urbanized and still yet rapidly urban urbanizing humanity of the Anthropocene? And how may a new, reinvented critical urban planning practice potentially contribute towards such shifts? Because, for certain, the questions raised by Marcuse in relation to Lefevbre's famous slogan 'the Right to the City' – '*whose right, what right and to what city?*' (Marcuse, 2009b: 185, emphasis added) – are still, and will forever remain, highly pertinent. But I would argue that we perhaps need to work towards expanding the range of possible answers that can be given to those questions far beyond what Marcuse appears to imagine. Particularly, it demands that we productively make use of the thought-space (and hence also action-space) opened up by these questions to address the relationally complex entanglements of urban commons, spanning across both geographical and ontological boundaries.

I will argue that this would demand that planners break out of the narrowly humanist ontological underpinnings of established planning theory, whether 'critical' or not, to instead develop methods that generate affordances for planning processes that do not only reproduce the same old well-trodden paths, which we by now know probably will lead us towards self-inflicted doom as a species. Instead, we urgently need to develop methods that enable planning processes to function as arenas that may contribute to engendering a new relationship to the world, and through which we hopefully might find ways to begin to deal with how we, as a species, have always been more-than-human. Such a practice would further entail an acceptance of responsibility for its own consequences and effects in relation to concrete and specific situations or episodes of urban planning and development work.

Nevertheless, for planning practice to be able to live up to its promise of becoming such a transformatory practice it must shed its modernist underpinnings and instead reinvent itself as a method for what Noortje Marres (2013) has called 'experimental ontology' towards a more ecological sensibility, utilizing visioning practices to open up lines of flight by asking questions of what may become of urbanity in the future in a way that can also make a difference in the present. Towards the end of the chapter I will return to Marcuse's proposition for developing a 'commons planning' as a potentially important development of planning practice and for caring for the urban commons in the Anthropocene. But I will argue that its potential to become so is not only dependent on our ability to find ways to expand the scope of what constitutes urban commons, but critically also demands a new, much more thoroughly relational understanding of the wider collectivity of urban commoners – those beings and existences whose fates are today tied to the future paths of development of cities and urban areas around the world.

Urban commons in a more-than-human world

Based on the by-now classical proposed alternatives for how to manage commons – *regulation, privatization* or *common-pool resource management* (CPRM) – since the mid-2000s scholars from a variety of disciplines have suggested various methods for dealing with specifically *urban* commons. For instance, there is the body of literature that conceptualizes urban public space as commons-type collective resources, where liberally inclined planning scholars such as Chris Webster and associates (e.g. Lee and Webster, 2006) have mobilized Garett Hardin's reasoning regarding the 'tragedy of the commons' (Hardin, 1968) to argue for property rights and privatization of urban public spaces. In response to this, law scholars such as Garnett (2012) and Foster (2006) have developed lines of argumentation that take a more ambivalent or sanguine approach regarding the potential of collective management of urban public space, primarily drawing upon Nobel laureate Ellinor Ostrom (e.g. Ostrom, 1990).

In addition, ecologists such as Johan Colding (2012) and Thomas Elmqvist at Stockholm University (e.g. Borgström et al., 2006), have found inspiration in Ostrom's CPRM to make a counter-argument about the potential ecological gains of collective management of urban green spaces through, for instance, community gardens, allotments and other similar institutional set-ups. Further, there are also the activism-inclined radical urban theorists who discuss urban commons in the form of more or less formalized cooperative urban communes that generate common resource pools (see, e.g. Gibson-Graham, 2011).

Taking a similar interest not only in the management of existing urban resources but also in understanding their production and becoming, urban sociologists Parker and Johansson (2011) focus on difficult-to-grasp urban amenities of a commons-like type – urban 'atmospheric' properties which are sometimes somewhat feebly conceptualized as, for instance, 'social capital', 'attractiveness' and so-called 'Jacobs' externalities' or 'buzz'. Focusing particularly on the latter, the production of the urban milieu as a commons, Kornberger and Borch (2015: 7) note that 'the notion of a commons as a self-evident and independent object makes little sense when applied to the urban. In the city, the commons is an inherently relational phenomenon.'

Taking this interest in the relational constitution of urban commons one step further, it also becomes relevant to develop a thinking that grasps 'the urban' in itself as a commons generating affordances for particular forms of life, always with its own wider milieu and conditions of existence; there through also recognizing that urbanity is – and has always been – fundamentally dependent upon a vast hinterland. This hinterland is to a certain extent directly traceable in a network form – for example, conceptualized as networks of urban metabolism – but also consists of global atmospheric commons on a planetary scale, the maintenance of which in turn is dependent on the well-being of innumerable other-than-human species, widely dispersed across the globe. Which means that urbanity is fundamentally dependent on its generally unrecognized 'absent Others' (cf. Law, 2004).

So, relating to the (in)famous problem of the commons free-rider, even if undomesticated animals and uncultivated plants generally have been considered 'free-riders' in the City (see also further below), it may be an important corrective to concomitantly recognize that if we shift the focus of analysis somewhat, we may come to recognize that cities to some degree also are free-riders on the planet's biosphere, and generally quite destructively so.

And this is where we begin to encounter some serious problems with all the previous work on urban commons that draws on the canon of the commons management literature. For one of the great taken-for-granteds of both Garrett Hardin's and Ellinor Ostrom's conceptualizations of the commons, as well as that of their disciples, is their clear and stable differentiation of commons and commoners, building upon a foundational view of the world in which humans are extractive actors and everyone or everything else are passive resources only acted upon. For instance, as noted in a paper by Jonas Bylund and Fred Saunders, the operationalization of Ostromian common pool resource theory is all about '*human* use of *natural* resources' (Saunders & Bylund, 2009: 3). Humans on one side, everything else on the other – a strict ontological divide between human and non-human, commoner and common, agent and structure, extractor and resource, culture and nature, subject and object, active user and passively used.

But if we allow ourselves to be inspired by scholars such as Bruno Latour and Donna Haraway to examine these seemingly neat and mutually exclusive categories a little closer, we may begin to see that they are not so neatly separable and that any actual worldly occurrence always consists of complex entanglements of elements, phenomena and tendencies that on paper may seem neatly separated from each other, but which, as they occur in the world, always turn up as irrevocably entangled and therefore end up messing up all these seemingly neat categories. As Latour (1993) has observed concerning the 'modern Constitution', the peculiar Western programmatic ontological separation of 'things Natural' and 'things Cultural', this was only ever an ostensive separation, for in practice links between these categories always proliferated covertly, leading him to the conclusion that '[t]here are only natures-cultures, and these offer the only possible basis for comparison' (Latour, 1993: 104). This obviously even goes to the heart of what it means to be human, the problematic bundle of tangled cultural and biological relations that in various ways have come to be categorized as the essence of mankind, when we take into account the fact that the human genome is found in no more than 10 per cent of the cells constituting a human body, while the remaining 90 per cent are made up of bacteria, fungi, protists, etc. (Haraway 2008: 2). So not only have we, with Latour, 'never been Modern', we have, further, 'Never been human', if we insist on defining humanity as a mode of being hermetically sealed off from and standing above other forms of life and existence (Haraway, 2008: 305).

The existence of urban commons must thus critically be grasped as relationally dependent on a host of 'others', from the planetary scale to the microscopic, spanning far beyond what is generally recognized. If we care for urban commons, we thus also need to bring these crucial prerequisites for their existence into the scope of our concern. In contrast to the exclusively and exclusionary human focus of mainstream and critical contemporary urban theory, such a thoroughly relational understanding of urbanity would thus instead proceed from Darwin's insight that all life on earth, urban or not, may well be knit together in an 'inextricable web of affinities' (Darwin, 1859: 434).

In what light, then, does the existing mainstream of the urban commons literature appear if we approach it with these glasses on? Take, for instance, the work of Colding (2012 citing McKay, 2000), who makes quite a convincing argument that the (urban) commoners of the past may be gainfully reconceptualized as the (urban) stakeholders of tomorrow. Even though Colding's text is appealing in its main argument, I believe that he makes two crucial, but in my view unhelpful, assumptions: first, that the commoners-cum-stakeholders are exclusively human; and second, that they are local, in a simple Euclidean sense.

I will go through these two problematic assumptions in turn, but first it is important to point out that we shouldn't be too hard on Colding for making them in the first place, seeing that they (especially the first one) are buttressed by more than 2000 years of 'Western' philosophy in general, and urban thinking in particular. This is a tradition which has repeatedly imagined the City (capital C) as the ideally exclusive dwelling of humans, standing in direct contrast to the savage nature imagined to exist outside of the city walls –walls that both physically and symbolically have been seen as generating a protective space in which the unique and supposedly superior traits that have been thought to distinguish humans from animals could be cultivated and fostered.

Relatedly, throughout the history of the global North, urban management practices have to a large extent aimed to generate the City as the antithesis of countryside and 'nature' through purging the 'beastly' and 'natural' aspects from the spaces and materials it works on, thereby contributing to separating 'nature from the city, both conceptually and materially' (Swyngedouw & Kaïka, 2008: 574). And, as further observed by Wolch et al. (1995: 735) '[t]he ideals of urbanization were based on a notion of progress rooted in the conquest of nature by culture', leading to a 'splitting apart of the urban and the rural as distinctive entities conceptually associated with particular human activities and attributes . . . the industrial and civilised city, the agricultural and barbarian countryside' (Philo, 1995: 666 – further referencing Williams, 1973). This is a process described by Atkins (2012) as 'the Great Separation' of urbanity and rurality that was enforced in nineteenth-century Western Europe – the culmination of the two-millennia-long, ever-expanded project of evacuating the presence of living animals from cities (see e.g. Atkins, 2012).

But, to begin with – focusing on but one category of non-humans – even if 'the animal' is still not recognized as a resident of 'the City' in the conceptual realm of Western philosophy and culture, myriads of other-than-human living things still have cities as their permanent homes: foxes in London, deer in Stockholm, boar in Berlin, raccoons all over the US, possums in Australia, peregrine falcons in Birmingham and rats, rabbits, pigeons, geese, dogs, bats, insects – the list goes on and on – all over the world. And we may therefore do well to take our cue from Tryggestad et al. (2013) to question Colding's strong assumptions by asking: 'what if the stakeholder is a frog?', and with this begin to unsettle the taken-for-granted assumption that the circle of stakeholders to be taken into account in any situation or context is exclusively composed of humans.

Pushing the argumentation further ahead, it is nevertheless also important to recognize that the path of reasoning being cleared here does not just hold an injunction to recognize that there are myriads of other-than-human urban denizens that hold a legitimate claim to life, liberty and the pursuit of happiness in urban environments. It also brings this insight together with a questioning of how we geographically trace the effects and dependencies of urban developments. Spatial planning processes always have to set boundaries as to what constitutes 'here', that which belongs to 'this place'— which is concomitantly enacted as worthy of care, thus normatively enacting spaces of care of an unavoidably exclusionary nature, and thereby enacting particular 'geographies of responsibility' (Massey, 2004) in practice. Unfortunately, much ongoing contemporary spatial planning practice lacks even the faintest self-reflexivity concerning how its practices contribute to enacting a 'we' that, whether recognized as such or not, nevertheless is always relationally constituted and dependent upon links that stretch far beyond ourselves in both time and space, to places and beings wholly Other than what we see as that which constitutes our proper selves (see also Healey, 2010; Massey, 2004). And since the geographical extent of responsibility and care is never given, but rather always assumed or decided, the geographical extension of 'stakeholderness' in relation to a particular episode of planning is always an issue that will be resolved by active boundary-drawing, whether recognized as such or not (cf. Metzger, 2013a).

Bringing together these two points, we may reflect that maybe the stakeholder is not just potentially a frog, but perhaps even a frog living on the other side of the planet, that could be considered to be affected in one way or another by the consequences of our 'local' urban development activities, if not only by the adverse consequences on the planetary atmosphere of, for example, car-oriented urban development schemes;[2] a frog that may hypothetically well be a key species in the maintenance of a crucial oxygen-producing ecosystem in the tropical rainforest. The combination of a thoroughly relational view of the geographies of planning and urban development together with a more-than-human sensibility thus demands a recognition of how humans, as a broadly urbanized species, are fundamentally implicated with, and largely dependent upon the fate of, myriads of other-than-humans on both the smallest of micro- and the widest of planetary scales: beings and existences that therefore deserve to be taken into our concern, if for no other reason, then for our own self-interest and self-preservation.

If we are not of the sanguine belief that such insights will grow somehow 'naturally' from 'the inside', it then becomes a rather urgent task to begin to think how we technologically, in the broadest sense, can begin to produce the apparatuses of engagement which may help to sensitize us to a geographically complex more-than-human world of always asymmetric becomings demanding critical responsibility (cf. Haraway, 2008). This is all about the messy business of living together in irrevocable intertwinement and partial co-dependence with things we consider other-than-ourselves, and of finding ways to attune to these conditions. And this is where I again turn to the prospects of 'commons planning'.

Urban commons planning, rehashed

How can we then find means to link our understanding and appreciation of localized more-than-human urban commons with a broader attention to their relational interdependencies with global atmospheric and biospheric commons? How can we develop practices that function as operationalizations of Isabelle Stengers' (2005) injunction to collectivize the nagging but today generally individual and introverted question 'what am *I* busy doing?' into a collective open questioning of 'what are *we* busy doing?', particularly in relation to urban development issues. Here, I would like to suggest that practices for critical urban planning in their contemporary 'spatial planning' variant (i.e. in the guise of long-range cross-sectoral urban policy development, as opposed to more narrowly defined 'urban design') might offer interesting potentials. Spatial planning is, broadly conceived, a planning practice that aspires towards identifying what must be done in the present for places to become somehow normatively 'better' in the future (cf. Healey, 2010) – a practice which is not only collective, but which could also potentially contribute to *collectivizing* concerns and caring for the fate of urban *milieux* through practices of common visioning and agenda setting.

It may spontaneously appear as odd to propose urban spatial planning as a means to achieve such a shift. Historically, the professional ideology of planning has been underwritten by a celebration of human intellect and agency, and in its high-modernist guise often appears as something of an apex of human hubris. But nevertheless, along the lines of Latour's (1993) argumentation, it must nevertheless be recognized that – even if its ideology and rhetoric has failed to recognize this – as a *practice*, planning has always had as a central concern and interest the deep entanglements of humans and non-humans and the agency produced in human/non-human interactions. Consequently, much of its focus has also centred on exploring the connections, mutual affectations and collective becomings of entities and phenomena which are otherwise generally slotted into ontologically divergent categories such as the 'natural', 'cultural', 'social', 'economic', 'material', 'symbolic', and so on (see further Metzger, 2013b).

As a consequence, throughout its history, planning has in practice repeatedly worked with and on non-humans to generate new forms of future-oriented more-than-human collectivities. Recognizing this, Jonathan Murdoch has argued that spatial planning could have the potential to, and also indeed *should*, be reconceptualized as a form of 'green governmentality', a crucial technology for organizing a more ecological human sensibility and society by way of paying attention to spatial patterns and entanglements across the human/non-human divide (2006: 155). I broadly agree with this and, elsewhere (Metzger, 2014a), I have proposed that Patsy Healey's (2010) conceptualization of the 'planning project' or 'place governance with a planning inclination' could be a productive starting place for such a practice.

The 'planning project', in Healey's guise, appears as a 'matter of care' combining a 'worry and thoughtfulness about an issue as well as the de facto belonging of those "affected" by it' with a 'strong sense of attachment and commitment' (de la Bellacasa, 2011: 89–90). By setting a group of disparate but proximate existences on a collective trajectory towards the future it assembles and enacts them as a specific type of collective thing, a 'place' in the form of a nexus of complex spatial entanglements containing interdependencies and obligations with other beings, near and far in both space and time (cf. also Metzger, 2014b). Crucially, for Healey 'the planning project' 'is not only about creating better opportunities and chances for individual people, or particular social groups, or people in a particular place, or even just the human species as a whole'; rather, '[i]t is about how we relate to all others who inhabit the world with us and to the broader natural forces that shape our planetary existence. Such a project demands that we try, in thinking about and acting with respect to place management and development, to see the larger issues in small actions and the little implications of greater endeavours' (Healey, 2010: 226).

Admittedly, labelling the type of practice proposed by Healey as a form of 'planning' generates the risk of conjuring up the wrong sort of associations of 'command and control'. For if there is but one thing we can be certain of regarding the future, it is that it will not turn out at all as we expected. What nevertheless makes spatial planning a promising practice for the enactment of a more-than-human urban sensibility here and now is that even though it in many ways is deeply speculative and thoroughly future oriented (as well as fundamentally reductionist, a point I will return to further below), it is nonetheless an activity that takes place in the present, which further aims at guiding action in the near future. As such – through visioning futures – it also articulates, and potentially rearticulates, 'society' in the present through enacting existing ontologies, norms, values and hierarchies – often reaffirming existing patterns and relations, but sometimes – wittingly and unwittingly – also destabilizing them (cf. Metzger, 2011, 2013b).

The crux of the matter of Healey's argument is that effective planning processes don't come about by themselves. They require active cultivation through skills, tools and technologies. And it is here that critical planning theory in general and 'commons planning' in particular, as discussed by Marcuse, can function as an important source of inspiration (Marcuse, 2009a). In his presentation of commons planning, Marcuse in effect argues that planners should no longer be satisfied with only applying Band-Aids on fundamental injustices, but instead dare to tackle the more profound challenges by opening up for discussions concerning 'what a city should be and for whom':

> Critically, the questions might start by asking: What is the purpose of public action in a particular case? Is it simply to find the highest and best use for a piece of land, or to raise tax revenues for the city, or to promote one business activity over another? Or is it to serve the common good, to improve the lives of individuals that are now or might potentially be affected by the public action?
>
> (Marcuse, 2009a: 101)

To some degree it thus appears as if Marcuse suggests that planners should utilize specific, singular planning processes as 'occasions for democracy' (Marres, 2005; cf. Metzger et al., 2014) in which 'a given issue becomes not one of simple land use, but of the goals of public action' (Marcuse, 2009a: xx), further enacting planning processes that engender planners to 'do the best they can within the context of real possibilities' in the singular specific case, but nevertheless doing so while at the same time 'seeking approaches that raise the larger issues as well', thus 'recognizing the limitations of planning, but not surrendering to them' (Marcuse, 2009a: 101).

Elsewhere, Marcuse has also conceptualized this practice (then labelled as 'critical planning') as consisting of the three consecutive steps of seeking to *expose, propose* and *politicize* urban development in the planning process (Marcuse, 2009b). In relation to this, he further suggests that focusing on 'Who benefits and who suffers' is 'an important part of planning analysis and can be eye-opening':

> Commons Planning would look at all groups having an interest, actual or potential, in the outcome (potential here is very important, for the standard 'stake-holders' definition of those needing to be involved rarely considers them), and ask what arrangement for the use of this particular site would best serve the interests of those groups, and how conflicting interests might best be solved.
>
> (Marcuse, 2009a: 99)

Further specifying his methodological proposal for planning, Marcuse then suggests that

> Above all, Commons Planning would then ask: What exercise of power is involved in creating the situation to be dealt with, and who is exercising that power? At whose expense is it being exercised, and what potential strength can be mobilized against that exercise?
>
> (Marcuse, 2009a: 99)

Marcuse is here elaborating upon quite an old-fashioned conception of 'power' as something inherently destructive and oppressive, but notwithstanding this I can certainly sympathize with his overarching ambition. The problem is that he makes numerous deeply problematic assumptions. To begin with, throughout his text, he appears to take for granted that there exists some given singular and indisputably optimal 'common good', even though the most casual examination of claims to the common good will inevitably show that any claim made on behalf of this elusive entity will always be an act of self-empowerment which, by necessity, is conditioned upon innumerable acts of translation. Articulations of the 'common good' are therefore probably better understood as 'risky' (because they are fallible and disputable) authority-enacting propositions through which specific actors attempt to position themselves as legitimate spokespersons for a particular collective. Thus understood, rather than being seen as some form of analytic uncovering of a pre-ordained truth, speech made in the name of the 'common good', then, instead appears as statements which (directly or indirectly) sketch the outlines of a specific collective, while concomitantly sketching a supposedly desirable trajectory towards the future for this particular grouping or entity (see further, e.g., Latour 2003 and 2004).

Second, a point I will only touch on briefly here (but see Metzger, 2013a for an elaborated discussion), problematically, Marcuse further assumes that 'all groups . . . actual or potential' that have 'an interest' in an issue can somehow exhaustively and 'objectively' be identified, even though the question of who is to be considered a legitimate stakeholder with a recognized interest in a particular planning issue often amounts to the central bone of contention in many planning processes. Finally, and perhaps most importantly in the present context, he clearly assumes that the 'individuals' and 'groups' who could have an 'interest' in planning processes are all exclusively human. But what about the interests of the urban denizens that fall outside of this quite narrowly framed category? For if we do not accept any kind of strict, cosmic dividing line between the human and other-than-human, that is, a more-than-human perspective, it becomes pertinent to ask whether there might be other types of 'individuals', 'groups' and 'interests' that could have an interest in the development of urban areas, also on the other side of the often taken-for-granted human/non-human divide.

All these three points of contention in one way or another boil down to the question of exclusions, and the drawing of boundaries of care and responsibility, in a way that goes well beyond discussions that wittingly or not treat various 'scales' as ontologically given. If we recognize the wicked relational complexity of this world, and how these relational webs of becoming constantly criss-cross the oft-imagined human/non-human divide, there is suddenly no self-evident place to draw the line between what is the 'us' and the 'them', the 'here' and the 'there', the affected and the irrelevant, the consultable and the dumb, the care-worthy and the worthless. All these are things that have to be worked out, in process, and taken responsibility for. Perhaps particularly the latter, for, as Lee and Stenner (1999) have noted in a different context, collectivity must always find its basis in 'radical exclusion of the "victim" from the benefits of membership' and the daunting ethico-political question then becomes to account for the victim, to take responsibility for these exclusions and to never give up asking: 'Who pays? Can we pay them back?' (Lee & Stenner, 1999).

So, how can we begin to find ways to work such considerations into planning processes? Ways to develop methods and devices for *slowing down* (Stengers, 2005) planning processes in a manner that allows for a painstaking working-through of these challenges and conundrums, which are in themselves unsolvable in abstraction but always can be concretely actualized in relation to specific events and episodes of planning? These would then entail methods that generate affordances for, or even *prompt* participants to perform, planning in ways that 'produce visions that "cut" differently the shape of a thing' through 'not only detecting what is there, what is given in the thing we are studying, but also think about what is not included in it and about what this thing could become – for instance if other participants were gathered by/in it', as a matter of collective care for that 'thing' (i.e. city, neighbourhood, place, etc.) (de la Bellacasa, 2011: 96, see further also Metzger, 2011).

Such methods would, for instance, do well to enact a planning process that not only prompted reflection on what and whom 'we' choose to plan *with* (the 'planning assemblage'), but also whom we plan *on* ('the planned assemblage') and *for* ('the assemblage planned for'). Another way to put this would be to say that there is need for critical reflection on the *agency* of planning (in all the polysemous meanings of that term), the *object* of planning and the *subject* (-to-become) of planning, as well as all the complex connections, intermixings and overlaps between these three groupings.

With regard to the agency of planning (who we plan *with*), the question regarding who should have a right to a say and influence over planning processes is a mainstay of contemporary planning theory (see Metzger, 2013a). Nevertheless, it is generally assumed that these questions of inclusion/exclusion pertain only to different groups of humans. But if we do not accept as self-evident that other-than-humans could not have some form of opinion or preference regarding the development of urban environments, we may do well to also note that the potential to actively consult non-humans in planning processes is not some fairytale pipe-dream but, rather, partly an ethical, but more broadly a plain technical, challenge (see further Metzger, 2014c), which nevertheless requires that we no longer take for granted that 'we can't know what they want, and neither do they know it'. For, as any dog owner or parent knows, there are signals there to read if only we take the time and go through the pain and effort to attempt to do so. So we should perhaps no longer ask ourselves the question, 'if nature could speak, what would it tell us?' and instead come to realize that myriads of creatures and beings are speaking to us all the time – if we would just learn how to listen properly (see also Desprets, 2008). From such a perspective, the granting of a right to a 'voice' in the planning process for any being or entity seen as affected by our actions, no matter if human or not, becomes a practical problem to be creatively solved, rather than an unsurmountable philosophical impasse.

To move on to the question concerning whom we plan *on*, this is broadly meant to engender reflection concerning with what right and legitimacy 'we' intervene in the unfolding fate of a specific entity or existence, and further, who or what will be the likely victims of or adversely affected by our decisions and actions. But to a large extent, the two final questions (as well as the first, actually) must be confronted as fundamentally interconnected, for they both concern how we enact boundaries between an 'us' and a 'them', between who and what is to be treated as a mere 'object' of action and who is worthy to be considered as a 'subject', and through this, who gets enacted as a means and who as an end.

In relation to a rehashed urban commons planning, the above considerations would have as their goal the enactment of urban planning processes in which we would never cease to ask, with Lee and Stenner, who pays, and whether we can 'pay them back'. This would be a process that continuously brings attention to how the boundaries of the commons, as well as the definition of the common-ers, are defined and patrolled (Borch & Kornberger, 2015). It would be a pro-cess which incessantly works towards bridging the destructive human/non-human divide in our enactment of the urban commons and the urban commoners, and which also further constantly articulates how the often seemingly local phenom-ena of urban commons are always-already irrevocably entangled into the fate and faring of planetary atmospheric and biospheric commons (see also Metzger & Rader Olsson, 2013), thus making it an urgent necessity to not only recognize that boundaries sometimes must be drawn, and to assume responsibility for how we decide to draw them, but also to recognize that we must always be conscious of the effects of these practices, and prepared to constantly challenge and reconsider them (cf. also Wolfe cited in Broglio, 2013: 181 and Latour, 2004).

Concluding discussion

So, to finally return to Marcuse's proposed three-step programme for critical urban commons planning – *expose, propose, politicize* – an alternative, more-than-human recomposition of this ambition could perhaps instead be phrased as something along the lines of an injunction to also *care, cosmopoliticize, and assume responsibility*. *Care* – here not understood as necessarily consisting of warm and tender loving, but rather more broadly as an ambition and ability to '"see" and to "hear" needs and to take responsibility for these needs being met' (Sevenhuijsen, 1998: 83). *Cosmopoliticize* – as in making cosmopolitical, i.e. continuously raising questions regarding the often radical (meta-) political con-sequences of decisions concerning what we recognize as real and existing, and what we discard as non-existent or irrelevant; and to experiment with what could potentially become if we were to open up for thinking differently around this (cf. Latour, 2004; Stengers, 2005). And finally, to *assume responsibility* for the perhaps necessary exclusions and Otherings that have to be part of any planning process aiming at generating some form of agency, and to do so even in the face of an almost unfathomably complex world in which it isn't always so easy – or rather it is impossible – to guess beforehand the extended effects of any interven-tion beyond the directly obvious. This thus involves constantly demanding 'us', however we define this, to constantly ask ourselves: can we look the likely victims of our actions in the eye and accept responsibility for the probable effects of our actions, while still maintaining that at this time and place, given the known status of things, this is – here and now – the correct decision to make?

From one perspective, there might not be that much radically new to this proposition for more thoroughly self-reflexive planning processes in the light of earlier developments of planning theory and methodology, as it is to some extent but an expansion upon the insights already touched upon by, for example, Schon regarding reflective practice (Schon, 1983) and Healey regarding communicative practices in planning (Healey, 1992). But what it does contribute is a further development of this programme of action through clearing out some of the more than two-millennia-old baggage of Western metaphysics that it is based on, as well as reproduces, untenable assumptions about the world we exist in, and instead allows for an opening up of new vistas for planning, hopefully leading the way towards a new, more-than-human planning sensibility more befitting to the Anthropocene (see further also Metzger, 2014a, 2014b, 2014c and 2015). This would be a planning sensibility that did not base itself on hubristic notions of human exceptionalism, but instead leaned against an extended, relational and situated ethics of care, nurturing and killing, which in turn rests upon incessant considerations and constantly ongoing negotiations concerning the composition and fate of an always-already entangled more-than-human urban *collectif* (cf. Callon & Law, 1995).

When the chips are down, planning is nothing less than a matter of life and death, perhaps even of war and peace.[3] If we begin to assume the responsibilities that come with engaging in such serious business we will most probably soon find that there is no ultimate and painless way to go about these things, no 'one size fits all' of planning methodology, but that the real challenge is about learning to act with responsibility and respect in connection with always-asymmetrical relations of living and dying, nurturing and killing. In the light of this, I do agree with Marcuse's (2009b: 194) 'rejection of the idea that the most desirable future can be spelled out, designed, defined, now, in advance, except in the most broad principles' and his powerful assertion that 'only in the experience of getting there, in the democratic decisions that accompany the process, can a better future be formed. It is not for lack of imagination or inadequate attention or failing thought that no more concrete picture is presented, but because, precisely, the direction for actions in the future should not be preempted, but left to the democratic experience of those in fact implementing the vision.' For, as Latour (2010) argues, the common world remains to be composed. And we may do well to add: over and over again. Because, as has been noted so many times, the default outcome of any episode of planning is obvious – and often monumental – failure, at least as seen in the gloomy light of afterthought. But, taking our cue from Samuel Beckett, the trick is of course to ever try again, and to constantly learn to fail better.

Notes

1 See e.g. Rockström et al. (2009), which particularly highlights ongoing biodiversity loss as one of the most alarming threats to human life conditions on a planetary scale.

2 This is thus a 'more-than-human' expansion to the discussion in Metzger (2013a), where I discuss the standard definition of 'stakeholderness' as encompassing those actors who are potentially affected by an action as being deeply problematic, since in a world marked by fundamental relational complexity, who is to be considered affected is never self-evidently given. I.e., should the citizens of the Maldives be considered as legitimate stakeholders in the planning of any and every city around the world, and thus be given a right to voice in that process, seeing that they are affected by the potential sea-level rise resulting from global warming which may be the result of a development of road traffic infrastructure in any urban area? There is no given answer to this question, only decisions of inclusion and exclusion that, once brought into consideration, often have wickedly daunting consequences.

3 See Tesfahuney and Ek (2014) as well as Bruno Latour's as yet unpublished 2013 Gifford Lectures, 'Facing Gaia'.

References

Atkins, P. (ed.) (2012). *Animal cities: beastly urban histories*. Farnham: Ashgate.

Barnosky, A. D., Matzke, N., Tomiya, S., Wogan, G. O., Swartz, B., Quental, T. B., . . . & Ferrer, E. A. (2011). 'Has the Earth's sixth mass extinction already arrived?' *Nature*, 471(7336) 51–57.

Borgström, S. T., Elmqvist, T., Angelstaam, P. & Alfsen-Norodom, C. (2006). 'Scale mismatches in management of urban landscapes'. *Ecology and Society*, 11(2), 437–466.

Broglio, R. (2013). 'After animality, before the law: interview with Cary Wolfe'. *Angelaki*, 18(1), 181–189.

Callon, M. & Law, J. (1995). 'Agency and the hybrid "Collectif". *The South Atlantic Quarterly*, 94(2), 481–507.

Castree, N. (2003). 'Environmental issues: relational ontologies and hybrid politics'. *Progress in Human Geography*, 27(2), 203–211.

Colding, J. (2012). 'Creating incentives for increased public engagement in ecosystem management through urban commons'. In Boyd, E. & Folke, C. (eds) *Adapting institutions: governance, complexity, and social-ecological resilience*. Cambridge: Cambridge University Press, 101–124.

Darwin, C. R. (1859). *On the origin of species by means of natural selection, or the preservation of favoured races in the struggle for life*. London: John Murray. 1st edition, 1st issue.

de la Bellacasa, M. P. (2011). 'Matters of care in technoscience: assembling neglected things'. *Social Studies of Science*, 41(1), 85–106.

Despret, V. (2008). 'The becomings of subjectivity in animal worlds'. *Subjectivity*, 23(1), 123–139.

Foster, S. (2006). 'The city as an ecological space: social capital and urban land use'. *Notre Dame Law Review*, 82(2), 527–582.

Garnett, N. S. (2012). 'Managing the urban commons'. *University of Pennsylvania Law Review* 160 (7): 1995–2027.

Gibson-Graham, J. K. (2011). 'A feminist project of belonging for the Anthropocene'. *Gender, Place and Culture*, 18(1), 1–21.

Haraway, D. (2008). *When species meet*. Minneapolis: University of Minnesota Press.

Hardin, G. (1968). 'The tragedy of the commons'. *Science* 162, 1243–1248.

Healey, P. (1992). 'Planning through debate: the communicative turn in planning theory'. *Town Planning Review*, 63(2), 143.

Healey, P. (2010). *Making better places: the planning project in the twenty-first century*. Houndmills: Palgrave Macmillan.

Hinchliffe, S. & Whatmore, S. (2006). 'Living cities: towards a politics of conviviality'. *Science as Culture*, 15(2), 123–138.

Holmberg, T. (2013). 'Trans-species urban politics stories from a beach'. *Space and Culture*, 16(1), 28–42.

Kornberger, M. & Borch, C. (2015). 'Introduction: urban commons'. In C. Borch & M. Kornberger, *Urban commons: rethinking the city*. London: Routledge.

Latour, B. (1993). *We have never been modern*. Cambridge, MA: Harvard University Press.

Latour, B. (2003). 'What if we talked politics a little?' *Contemporary Political Theory*, 2(2), 143–164.

Latour, B. (2004). *Politics of nature: how to bring the sciences into democracy*. Cambridge, MA: Harvard University Press.

Law, J. (2004). *After method: mess in social science research*. London: Routledge.

Lee, N. & Stenner, P. (1999). 'Who pays? Can we pay them back?' In Law, J. & Hassard, J. (eds), *Actor network theory and after*. Oxford: Blackwell.

Lee, S. & Webster, C. (2006). 'Enclosure of the urban commons'. *GeoJournal*, 66(1), 27–42.

Marcuse, P. (2009a). 'From justice planning to commons planning'. In Marcuse, P. et al. (eds) *Searching for the just city*. New York: Routledge, pp. 91–102.

Marcuse, P. (2009b). 'From critical urban theory to the right to the city'. *City*, 13(2–3), 185–197.

Marres, N. S. (2005). 'No issue, no public: democratic deficits after the displacement of politics'. Doctoral thesis. Amsterdam: Department of Philosophy, University of Amsterdam.

Marres, N. (2013). 'Why political ontology must be experimentalized: on eco-show homes as devices of participation'. *Social Studies of Science*, 43(3), 417–443.

Massey, D. (2004). 'Geographies of responsibility'. *Geografiska Annaler: Series B, Human Geography*, 86(1), 5–18.

Metzger, J. (2011). 'Strange spaces: a rationale for bringing art and artists into the planning process'. *Planning Theory*, 10(3), 213–238.

Metzger, J. (2013a). 'Placing the stakes: the enactment of territorial stakeholders in planning processes'. *Enviroment and Planning A*, 45(4), 781–796.

Metzger, J. (2013b). 'Raising the regional Leviathan: a relational-materialist conceptualization of regions-in-becoming as publics-in-stabilization'. *International Journal of Urban and Regional Research*, 37 (4), 1368–1395.

Metzger, J. (2014a). 'Spatial planning and/as caring for more-than-human place'. *Environment & Planning A*, 46(5), 1001–1011.

Metzger, J. (2014b) 'The subject of place: staying with the trouble'. In Haas, T. & Olsson, K. (eds) *Emergent urbanism*. Aldershot: Ashgate.

Metzger, J. (2014c). 'The moose are protesting: conceptualizing planning politics across the human/non-human divide'. In Metzger, J., Allmendinger, P. & Oosterlynck, S. (eds) *Planning against the political: democratic deficits in European territorial governance*. New York: Routledge.

Metzger, J. (2015). 'The city is not a Menschenpark: conceptualizing the urban commons across the human/non-human divide'. In Borch, C. & Kornberger, M. (eds) *Urban commons: rethinking the city*, London: Routledge.

Metzger, J. & Rader Olsson, A. (2013). 'Introduction: the greenest city?' In Metzger, J. & Rader Olsson, A. (eds) *Sustainable Stockholm: exploring urban sustainable development in Europe's greenest city*. New York: Routledge, pp. 1–9.

Metzger, J., Allmendinger, P. & Oosterlynck, S. (2014). 'The contested terrain of European territorial governance: new perspectives on democratic deficits and political displacements'. In Metzger, J., Allmendinger, P. & Oosterlynck, S. (eds) *Planning against the political: democratic deficits in European territorial governance*. New York: Routledge, pp. 1–28.

Murdoch, J. (2006). *Post-structuralist geography: a guide to relational space*. London: Sage.

Ostrom, E. (1990). *Governing the commons: the evolution of institutions for collective action*. Cambridge, UK: Cambridge University Press.

Parker, P. & Johansson, M. (2011). 'The uses and abuses of Elinor Ostrom's concept of commons in urban theorizing'. Paper presented at International Conference of the European Urban Research Association (EURA) 2011 – Cities without Limits 23–25 June, Copenhagen. Available at: http://dspace.mah.se/dspace/bitstream/handle/2043/12212/EURA%20conf%20version3.pdf;jsessionid=B3F6A280E3DB378ED0A4F507EB0FA6DB?sequence=2.

Philo, C. (1995). 'Animals, geography, and the city: notes on inclusions and exclusions', *Environment & Planning D*, 13, 655–681.

Rockström, J. et al. (2009). 'A safe operating space for humanity'. *Nature*, 461(7263), 472–475.

Saunders, F. & Bylund, J. (2009). 'On the use of actor-network theory in a common pool resources project'. *The Commons Digest*, 8, 1–10.

Schön, D. A. (1983). *The reflective practitioner: how professionals think in action*. New York: Basic Books.

Sevenhuijsen, S. (1998). *Citizenship and the ethics of care: feminist considerations on justice, morality and politics*. London: Routledge.

Stengers, I. (2005). 'The cosmopolitical proposal'. In Latour, B. & Weibel, P. (eds) *Making things public: atmospheres of democracy*. Karlsruhe: Engelhardt & Bauer, pp. 994–1003.

Swyngedouw, E. & Kaïka, M. (2008). 'The environment of the city . . . or the urbanization of nature'. In Bridge, G. & Watson, S. (eds) *A companion to the city*. Blackwell: Oxford.

Tesfahuney, M. & Ek, R. (2014). 'Planning as war by other means'. In Metzger, J., Allmendinger, P. & Oosterlynck, S. (eds) *Planning against the political: democratic deficits in European territorial governance*. New York: Routledge.

Tryggestad, K., Jutesen, L. & Mouritsen, J. (2013). 'Project temporalities: how frogs can become stakeholders'. *International Journal of Managing Projects in Business* 6(1), 69–87.

Whatmore, S. (2002). *Hybrid geographies: natures, cultures, spaces*. London: SAGE.

Wolch, J. (2002). 'Anima urbis'. *Progress in Human Geography*, 26(6), 721–742.

Wolch, J. R., West, K. & Gaines, T. E. (1995). 'Transspecies urban theory', *Environment & Planning D*, 13, 735–760.

Zalasiewicz, J., Williams, M., Steffen, W. & Crutzen, P. (2010). 'The new world of the Anthropocene'. *Environmental Science & Technology*, 44(7), 2228–2231.

7 Occupy the future

Julian Brigstocke

Occupy the future

> Global capitalism is *foreclosing on our future* by tearing apart the fabric of our homes, communities, and ecosystems. While our resources go to bail out Wall Street, corporations are destabilizing our climate, fouling our waters, selling us food that makes us sick, and destroying the very life-giving systems that we depend on.
>
> (Movement Generation, 2011)

A remarkable feature of the Occupy movement has been the widespread calls for protesters to 'Occupy the Future', reclaiming time as a form of commons: something that is collectively practised, shared, and distributed. In one striking poster (Figure 7.1), a faceless businessman, coloured in red and with an angel's halo and a devil's tail, walks towards a little girl holding a large banner saying 'Occupy your future'. The girl stands in front of a crowd of protesters, and stands firm on the bottom edge of the image. The businessman, by contrast, despite his size, is lost in the middle of the image, anchorless in a sea of grey. The little girl, through her age and gender, embodies conventions of purity and reproductive futurity; the crowd behind her offer the strength to overcome the satanic corporate world. The image articulates a theological temporality of innocence, salvation, and fulfilment.

In this chapter, I examine the ways in which time – in particular, the future – can be conceptualized and 'occupied' as a form of commons. Exploring the aesthetic strategies through which time and the future are figured in posters, artworks, and advertisements that campaign for future justice (see also McKee, forthcoming), I highlight four key aesthetic figures of the temporal commons: foreclosure; obduracy; prefiguration; and future generations. Underlying each of these figures is an insistence on the plenitude, openness, and communal nature of an occupied future, in contrast to the isolated, austere, and privatised futures presaged by contemporary capitalism. Distinguishing myself from attempts to 'represent' occupied futures in the present, thus rendering them calculable and co-present, I will argue that the promise of the call to 'occupy the future' comes from an attunement to forms of temporality that recover a sense of the future as unknown, incalculable, but insisting within alternative practices in the present.

Figure 7.1 Occupy Your Future.

Source: Image designed by Hello Cool World, courtesy of TheCorporation.com.

The temporal register evident in the rhetoric of Occupy the Future, which insists on reclaiming the future as a commons and thereby opposing the individualisation and privatisation of the future, is a striking counterpoint – and theoretical challenge – to the largely spatial vocabularies and aesthetics of the commons. It does not represent an avant-gardist emphasis on leading an advance party towards a future that is known in advance, but a determination to figure or 'prefigure' alternative futures in the present (see de Angelis, 2007). In this chapter, then, I analyse some 'figures' of futurity in Occupy the Future and related campaigns for intergenerational justice.

Figuration is a complex aesthetic practice with roots in the semiotics of Western Christian realism and the temporalities of progress, fulfilment, apocalypse, and salvation that are associated with this tradition. Figures establish connections between events or persons so that a figure both signifies itself and a second, and this second involves the fulfilment of the first (Auerbach, 2003). Figuration has many intersecting meanings: figures are drawings, graphic representations, faces (as with the French 'figure'), calculations, and spatial arrangements in rhetoric. As Donna Haraway (2007: 11) writes, 'Figures do not have to be representational and mimetic, but they do have to be tropic; that is, they cannot be literal and self-identical. Figures must involve at least some kind of displacement that can trouble identifications and certainties. Figurations are performative images that can be inhabited. Verbal or visual, figurations can be condensed maps of contestable worlds.' Figures carry *authority*; they make claims that demand a response (Brigstocke, 2013; Millner, 2013). As Dawney (2013: 43) argues, invoking the 'figure' makes possible an interpretive move away from fixed categories such as body and subject, towards 'a way of conceptualizing the affective capacities that are held by figures that are both material and symbolic, that are produced by and produce the social'. The figure is 'part of a distributed set of relations and is constituted through affective forces that bring it to visibility' (Dawney, 2013: 43). Moreover, figures mobilize and draw their authority from specific spatialities and temporalities.

In what follows, I examine four material-semiotic figures which struggles for the temporal commons have mobilized and drawn authority from. We start with a discussion of the 'temporal commons', describing some ways in which we might consider the future to be something that is being enclosed, privatised, or individualised, and identifying practices that 'occupy' futures that are collective, shared, and hopeful. I move on to analyse four material-semiotic figures of the temporal commons: foreclosure; obduracy; prefiguration; and future generations. Figures of futurity that attempt to represent future generations in present political bodies, I argue, risk rendering future 'others' contemporary and co-present, and thereby performatively enact a future that is largely congruent with the present. Instead, the most powerful forms of figures of futurity make palpable modes of futurity that are radically distant and other: modes of 'time without me'.

Foreclosing on the temporal commons

In the opening paragraph of Occupy Wall Street's description of the movement, they state that their aim is 'to fight back against the richest 1% of people that are writing the rules of an unfair global economy that is *foreclosing on our future*' (Occupy Wall Street, 2015, emphasis added). Echoing this theme of foreclosure, on 20 January 2012, several thousand residents of the San Francisco Bay Area occupied the financial district as part of the Occupy Wall Street West actions, demanding that banks should put an end to a wave of predatory evictions and foreclosures. Protesters targeted banks and corporations, and disrupted the auction of foreclosed homes. They also occupied Citicorp's main office, staging a mock foreclosure, piling furniture, and moving boxes into the revolving door at the main entrance. In the same way as the movement reappropriated the meaning of 'occupation' away from its military and authoritarian connotations to one of liberation and sharing (Vasudevan, 2015), the Foreclose on Wall Street protests challenged the redistribution of wealth from the poor to the rich that the foreclosures enacted, and insisted that it was the banks themselves that had reneged on their promises and social liabilities. A great debt was owed by the banks to the people, not the other way round. The Foreclose on Wall Street protests aimed literally to discredit the banks and to call in the social debts owed by the banking sector to the '99%'. In doing so, they deployed an aesthetic figure of foreclosure that challenged the temporal logic through which ordinary people's futures could be happily destroyed in order to maintain temporal stability (business as usual) in the banking sector. Foreclose on Wall Street, then, challenged the temporal logic of financial capitalism, insisting on the debt owed by the '1%' to the '99%'. In doing so, the protesters raised the prospect of creating forms of temporal commons where futures are not privatised or owned by banks, but are cooperatively produced and fairly distributed.

The 'commons' refers to forms of human solidarity based on an ethos of sharing, cooperation, and generosity in opposition to individual egotism (see Introduction, this volume). Peter Linebaugh (2014) draws a helpful distinction between the commons (which stands in opposition to individualism) and the public (which stands in opposition to the private). In its classic historical formulation, the commons refers to the land shared by peasants to grow crops and raise animals, until such common fields were privatised and enclosed with hedgerows by landowners, with the legal sanction of the state. While struggles over common land remain widespread, the 'commons' now refers to a much wider range of tangible and intangible goods, resources and spaces. Practices of everyday commoning (Bresnihan & Byrne, 2014) create forms of egalitarian sociality that may be 'anti (against), despite (in) and post (beyond) capitalist' (Chatterton et al., 2013). Practices of commoning, however, are not imposed by hierarchical state bureaucracies, but require the active, unalienated participation of the commoners.

Political languages of the commons are saturated with spatial vocabularies. Dominant imaginaries of the commons, I would suggest, still remain tied to ideas of common land, physical enclosure, and spatial occupation. If we are to pay due regard to the fundamental geographical insight that spatial relations are inseparable from temporal relations, however, it is necessary to develop a fuller account of the temporal registers of a spatial politics of the commons (see de Angelis, 2007). Indeed, here I wish to argue, extending the arguments of Bluedorn and Waller (2006), for the need to include a full recognition of the *temporal* commons in wider struggles for the commons. Such temporal commons demand a political and aesthetic vocabulary that extends beyond (orthodox readings of) concepts of '*en*closure' and towards a vocabulary of temporal '*fore*closure' that challenges the ways in which dominant regimes of capital are privatising and individualising time, and not just space.

According to Bluedorn and Waller, the 'temporal commons' can be characterised as 'the shared conceptualisation of time and the set of resultant values, beliefs, and behaviours regarding time, as created and applied by members of a culture-carrying collectivity' (2006: 357). Time, they suggest, is a crucial aspect of the 'intangible commons'. However, the temporal commons are being enclosed through logics of efficiency that replace collective, shared time with privatised, individualised, and commodified time. One of the fundamental mechanisms through which the temporal commons are foreclosed is through the market's privatisation of time (making all time available for transactions; valuing time only for its transaction potential; ignoring the *quality* of time). As Bluedorn and Waller perceptively observe, this form of enclosure has received less-vigorous contestation than enclosures of more spatially defined commons.

This theory of the temporal commons, however, needs to be extended to recognise the ways in which the temporal commons, far from being inherently immaterial, are also a *material* aspect of social life. To presuppose that time is essentially immaterial (and thus, implicitly, that space is inherently material), is to reproduce a highly problematic privileging of time over space in social theory (see Massey, 2006). Rather, time is produced, just like space, as the effect of specific material practices. This is one reason why, unlike the immaterial commons, time is often governed by a logic of scarcity. Time is constantly reproduced through material spatial practices (May & Thrift, 2001). The temporal commons should not be limited to a 'shared conceptualisation of time', as if time in and of itself had no substantial reality, but should be seen as an essential component of all commons. The temporal commons, therefore, would be better characterised as the product of multiple, heterogeneous practices of time and the resultant values, beliefs, and affective experiences of time.

Several ways in which the temporal commons have been foreclosed through privatisation and individualisation can be remarked upon (Stavrides, 2013; Lejano & Ericson, 2005). The disciplinary privatisation of time perfected by the nineteenth-century factory, for example, is perhaps the most often-discussed example of the commodification of temporal relations (Thompson, 1967). This functions through the imposition of forms of 'clock time' that divided time into easily quantifiable (and hence marketable) units of measure: a logic that has achieved its most ludicrous example in the forms of 'high-frequency trading' where reductions of time delays of thousandths of a second in flows of information become worth billions of dollars. Whilst the very rich history of time keeping cannot be reduced to a single narrative of the modern imposition of clock time (see Glennie & Thrift, 2009), nevertheless the disciplinary ordering and regulating of time in spaces such as factories, barracks, hospitals, and workhouses was an important feature of the foreclosure of the temporal commons in industrial modernity. Thus, political struggles over time in the nineteenth century were usually focused on reducing the length of the working day, as well as on raising the minimum age of child labour. As these struggles achieved partial success, and a middle class emerged with significant periods of leisure time at weekends, holiday periods, and at the end of the working day, leisure time was also increasingly commodified.

A powerful poster of the Occupy movement gestures towards this kind of foreclosure of the temporal commons (Figure 7.2). A golden Sun rises over a crowd of Occupy Wall Street protesters. This Sun rises into two blank, angular spaces. Visible underneath it are the smudged outlines of a clock. A figure of solar abundance eclipses the disciplinary strictures of clock time, and washes the crowd below with a golden glow. The image heralds a new occupation of time itself. 'Whose time?', the poster asks. 'Our time. May 1 Rising'. The poster implies that 'our time' is not the commodified temporality of clock time. Rather, it is a time that is shared and practised communally, and stands in abundance rather than scarcity. Here, time is not subjected to calculation, but is celebrated in its potentiality, warmth, and plenitude.

A different kind of foreclosure of the temporal commons, however, can be discerned in Gilles Deleuze's diagnosis of the 'control society'. In his celebrated essay, Deleuze argues that contemporary societies are experiencing a rather different kind of privatisation of time. Whereas forms of disciplinary power in the eighteenth and nineteenth centuries initiated the 'organisation of vast spaces of

Figure 7.2 Whose time? Our time!

Source: Elizabeth Knafo & MPA. Creative Commons licence.

enclosure', the control societies of the present operate within a different kind of temporality. 'In the disciplinary societies one was always starting again (from school to the barracks, from the barracks to the factory), while in the societies of control one is never finished with anything' (Deleuze, 1992). These different modes of power presuppose different temporalities of justice and judgement: on the one hand, 'apparent acquittal' (between two incarcerations); on the other, 'limitless postponements', where judgement is always 'to come'. In control societies, time is not so much enclosed as it is controlled as an *open* system which, by continually postponing judgement, leaves people subject to continual regimes of evaluation, quality control, retraining, competition, and re-evaluation. This occupation of time effaces distinctions between work and leisure time, and occupies the future indefinitely, via an infinite deferral of judgement and justice.

For the Occupy movement, the way in which this control of the temporal commons has resonated most strongly is in relation to debt. As levels of household debt rise to historic highs, subjects tie themselves to strict temporal regimes of repayment and financial behaviour. As debts are privatised, outsourced, and speculated on, new forms of control become possible. Subjects are disciplined into regular calendrics of repayment (Guyer, 2007; Lazzarato, 2012), or, increasingly, debt itself becomes mobile and deferrable, controllable through identifying possibilities rather than calculating probabilities (Atkins, 2014; Amoore, 2013). Projected future profits, meanwhile, become a key source of wealth in the present.

For movements such as Strike Debt (an offshoot of Occupy), debt is a means of keeping people 'isolated, ashamed, and afraid'. 'Debt is a tie that binds the 99%. With stagnant wages, systemic unemployment, and public service cuts, we are forced to go into debt for the basic things in life — and thus surrender our futures to the banks . . . We want an economy in which our debts are to our friends, families, and communities — and not to the 1%' (Strike Debt, 2015). A poster for

Figure 7.3 Strike Debt.

Source: Strike Debt.

Strike Debt (Figure 7.3) shows a tiny figure in a vast, empty, homogeneous space (see also McKee, 2013). The figure is entirely isolated, hunched over in anxiety. The image dramatises the ways in which debt acts as a form of futurity that is individualised, privatised, and isolated. On the horizon, however, a Sun is once again visible, a dawn beginning to break over the desolate night (see Dean, 2012 for a powerful reading of the 'Communist horizon'). 'You are not a loan,' the poster declares. Again, solar generosity stands on the brink of overturning the night of isolating futures of indebtedness.

Finally, we can also point to the ways in which the growing awareness of climate change, ecosystem destruction, and mass species extinction is leading to a collective sense of foreboding concerning the nature of the world that we bequeath to the future. This results in futures that are foreclosed on in the most extreme sense, threatening not just devalued futures but *destroyed* futures: futures that contain only an ongoing repetition of violence, destruction, extinction, and the uneven geographies of climate injustice. Such ecological destruction can be viewed as a foreclosure of the future in the sense that it risks irreparably damaging the future for future inhabitants of the Earth. Such foreclosure is a direct consequence of the dramatic expansion of the capitalist economy over the last two hundred years, and its cycle of accelerating consumption, pollution, and destruction of resources, life, and beauty. This kind of privatisation of the future also has the effect of rendering the present spectral and monstrous. The unimaginably vast time-scales of geological time, with which we are increasingly confronted in our day-to-day lives, cannot be apprehended, and slip in and out of our consciousness and visibility (Morton, 2013).

A number of other ways in which time is being commodified, privatised, and individualised could be described here. The key point is that the temporal commons – forms of shared, collectively produced temporalities – are being individualised, privatised, and commodified. New ways of commoning time need to be – and *are* being – invented. Spaces of alternative trading such as 'time banks' or 'time currencies', for example, contest the reduction of time to measures of financial value (see Hughes, 2005). Time banks are community currencies which reward an hour of community volunteering with one credit of a time-based local currency. Members use earned time credits to purchase services from the scheme. Such schemes aim to redefine the socially excluded as providers of useful services, rather than passive recipients of help (Seyfang, 2003). They are based on a principle of sharing and valuing time equally, in contrast to the dominant system that renders some people's time extremely valuable and other people's time 'worthless'. If, in practice, time banking has been observed to suffer from imbalances of power and participation between the skilled and unskilled, with professional members of these initiatives withholding their skills in fear of losing income (Lee et al., 2004), nevertheless time banks aim in principle for systems of cooperation, sharing and mutual support that achieve a certain degree of distance from the capitalist economy.

In the struggles for the temporal commons associated with the Occupy movement, however, we see a practice of temporal commoning that moves beyond logics of equivalences and exchange, towards an ethics of generosity and abundance.

Occupying time

The diverse movements fighting for the temporal commons are creating a powerful living archive of future-oriented practices. Here, I wish to highlight two more key material-semiotic figures of the temporal commons, before going on to consider in more detail the distinctive role of the figure of 'future generations'.

First, a key way of challenging the temporalities of enclosure and privatisation has been through experimenting with different speeds of social practice. In resonance with practices such as the 'slow' movement (see Sharma, 2014), perhaps the most visible of these has been the Occupy movement's dramatisation of *obduracy* (i.e. stubborn persistence through time). In contrast with the conventional repertoire of protests and marches that take place over a short period of time and are soon forgotten (of which the vast and largely ineffective 2003 worldwide protests against the Iraq war were perhaps the nadir), the Occupy camps filled public spaces (and created new ones) with the aim of persisting for as long as possible, resisting the neo-liberal temporalities of continual flux and speed and insisting on the right to persist in the creation of new worlds. Judith Butler has linked this obduracy to the authority of physical bodies themselves, with the Occupy camps 'enacting the demands of the people through the gathering together of bodies in a relentlessly public, obdurate, persisting, activist struggle that seeks to break and remake our political world' (Butler, 2011b). Bodies on the street, by occupying space, repeating that occupation, and persisting in that occupation, pose the challenge to privatisation and enclosure in corporeal terms. The persistence of the body calls the legitimacy of the state into question, 'and does so precisely through a performativity of the body that crosses language without ever quite reducing to language . . . Where the legitimacy of the state is brought into question precisely by that way of appearing in public, the body itself exercises a right that is no right; in other words, it exercises a right that is being actively contested and destroyed by military force, and which, in its resistance to force, articulates its persistence, and its right to persistence' (Butler, 2011a). This obduracy has also created a remarkable space in which there is time for extensive political debate, discussion, and experimentation with new social architectures. Through obduracy and persistence in the face of hardship, violence, and imprisonment, a space is created in which time can be socialised and shared.

Second, the temporalities of 'prefigurative' politics have been central to the politics of Occupy and related anarchist movements. This emphasis on prefiguration can be seen as a response to a further mode of foreclosing the future, in which contemporary political culture struggles to articulate alternatives to contemporary structures of power. Frederic Jameson, for example, has long been describing the 'impossibility' of imagining the future in contemporary societies. As 'prisoners' of cultural and ideological enclosure, he argues, we are unable to grasp the social totality in which we are embedded, and thus unable to imagine or project beyond it (1982). Utopianism no longer makes designs on the future (and risks colonising its openness and otherness), but looks for the heterogeneity, diversity, and differences at work in the present moment, carving out spaces of alterity at work in the social world. Reticence towards closing down the future with totalising utopian projects risks leading to a different kind of privatisation of time. This form of enclosure is manifested in an inability to imagine social alternatives to a present that seems unassailable, which means that the future becomes individualised, imaginable only in relation to individual goals rather than wider social transformation. This again gives the present a spectral quality. As the experiential *depth* or 'four-dimensional' quality of time recedes, the present becomes increasingly unreal. It is inhabited by ghosts, incorporeal manifestations of barely remembered pasts and absent, empty futures, and monsters such as the 'vampire squid', that became a widespread motif in the aesthetics of Occupy.

Occupy's response to this has been to articulate prefigurative politics that live out forms of organisation and social relationships that reflect the future society they seek to create, actively creating a new society in the shell of the old (Ince, 2012; Springer, 2013: 408). The notion of prefigurative politics signifies 'the idea that the struggle for a different society must create that society through its forms of struggle' (Holloway 2010: 45). Prefiguration refers to 'a political action, practice, movement, moment or development in which certain political ideals are experimentally actualised in the "here and now", rather than hoped to be realised in a distant future. Thus, in prefigurative practices, the means applied are deemed to embody or "mirror" the ends one strives to realise' (van de Sande, 2013: 230). Prefiguration implies a folding of the future into the present, evading means/end distinctions and living alternative futures in the present through experimental forms of social practice and political organisation. Prefiguration involves the

Figure 7.4 The Happiest Place on Earth.

Source: David Shankbone. Creative Commons licence.

demand actively to 'be the change you want to see' (Figure 7.4), experimenting with new forms of living, organising, building, and thinking in doing so.

Representing future generations

Another notable feature of struggles over the temporal commons has been the frequency with which figures of 'future generations' are mobilised in order to combat the foreclosure of the future through the destruction of the environmental commons. Occupy London, for example, demanded the creation of a 'positive, sustainable economic system that benefits present and future generations' (Occupy London, 2011), a gesture that has been echoed across many different Occupy movements. Moreover, as the large-scale planetary issues raised by the concern over the environmental commons, and the ever-expanding temporal registers of social action, become ever more pressing, there has been a growing wider interest in asserting the claims of future generations upon the commons. The 2003 Tomales Bay Institute Report on *The State of the Commons*, for example, argues that 'the commons' embraces 'all the creations of nature and society that we inherit jointly and freely, and hold in trust for future generations'. The commons thus encompasses, it argues, future common assets, common property, and common wealth. Building on this definition of the commons as something that can be claimed by future generations, the Climate Legacy Initiative report written by commons activists and researchers at the Vermont Law School makes a number of legislative, regulatory, and judicial proposals. Recommendations include implementing environmental rights for future generations; passing Acts to define the environmental legacy that should be left to future generations; establishing offices of 'legal guardians' to act on behalf of future generations; and the United Nations' adopting a declaration formally recognising the atmosphere as a global 'commons' shared by present and future generations.

Campaigns such as these make strong cases for protecting the future as a commons through legal frameworks, recognising that the law has historically been the most powerful tool for protecting various forms of commons. In doing so, they raise the question of intergenerational justice: the extent to which unborn lives can make claims upon the living. In addressing this, there has been a growing chorus of calls for future generations to achieve some form of representation within formal political spheres. The UK's Alliance for Future Generations, for example, argues that 'Our civilisation has developed to the point where our actions impact not just other people and our local environment, but the whole planet and the conditions of life for centuries to come ... We stand to leave to future generations an impoverished common inheritance ... Our democracy has not yet caught up with the opening of our eyes to our global and long-term reach – we have not yet extended the vote to future generations.' Similarly, Earth Manifesto propose a Bill of Rights for Future Generations, arguing that: 'The most severely under-represented interests in our political system are young people under the age of 18, because they cannot vote, and every person to be born in the future ... [A] new Bill of Rights must be designed to ensure that future generations have reasonable prospects for freedom, dignity, prosperity, financial stability and survival ... Congress and nations worldwide should "pay forward" some good deeds to improve the prospects for our children and their descendants' (Earth Manifesto, 2015).

This way of figuring the temporal commons aligns with other figures of the temporal commons by articulating an ethos of generosity that rejects the logic of austerity and 'pays forward' to the future, imagining time in terms of a plenitude and abundance that is made possible by a radical redistribution of wealth away from the '1%' in the present. Where it diverges from the earlier figures, however, is in assuming that it is possible to 'represent' the needs, wishes, and desires of the future in the present. Moves to give future generations formal representation in present-day political bodies, through offices such as the 'ombudspersons' for future generations (or similar bodies) in Wales, Hungary, and Finland, stand in danger of entering into a performative contradiction. By assuming that it is possible to speak 'for' future generations, there is a risk of assuming that the future will look similar to the present, that the future is knowable and calculable. Yet this precludes both the possibility of a drastically altered future, whether one in which catastrophic climate change, extinction, or war have transformed the world and the people within it, or one in which radical political transformation has interrupted the world from its present course. By using the representation of future generations as way of calling for change (and averting undesirable change), it seems to implicitly assume a model where the future does *not* radically change. It performs the stability that it aims to undermine.

A 2014 advertisement for Seventh Generation, a US company that makes more sustainable diapers and cleaning products and campaigns for a 'toxin-free generation', encapsulates the ways in which drawing on the authority of future generations risks falling into a problematic hetero-normative domestication of the future's alterity or 'otherness' (Figure 7.5). The narrative of the advert, which is

Figure 7.5 Seventh Generation.

Source: Seventh Generation

presented as an animation, involves a mother writing a letter to her young daughter, to be opened on the day in the future when the daughter gives birth to her own child. As she writes the letter, her daughter is caring for a teddy bear in its bed. 'You're going to be a great parent, Katy. I can already tell', the mother writes, before giving advice about trusting her instincts, and mentioning her worries about toxic chemicals in the home. The advert ends with a message that, 'As parents, there's only so much we can do to protect future generations', and asking us (the 'we' within which it includes the viewer) to join the campaign against toxic chemicals. The advert invokes a highly normalised, sentimental temporality of organic, gradual process based on female heterosexual reproductivity, or what Lee Edelman (2004) calls 'reproductive futurism'. It presupposes an intensely gendered and hetero-normative vision of the future, one that combines pessimism about radical change ('there's only so much we can do') with an optimism in the power of forms of alternative consumption, which do not challenge wider social norms, to make small changes. Future generations are assumed to be those *closest* to us (the viewer's own children's children); they are made knowable, calculable, and easily assimilated into contemporary norms. 'Future generations' are invoked to make the future familiar and unthreatening. Care for the future is reduced to an individualised, privatised sense of care for one's own family.

An ethic of generosity and hospitality to the future, I am arguing, requires thinking of future lives as *distant* others, not as familiar people who have an organic reproductive relation with oneself. Intergenerational justice – just like spatial justice (see Soja, 2010) – has to be extended to distant others, not simply those closest to 'us'. To be clear, in making this argument I am not suggesting that invoking future generations as a source of authority for political claims is necessarily invalid. Rather, I am arguing that the fight for the temporal commons can be conducted in the present through generosity on behalf of a future that is unknown, mysterious, and strange, rather than by domesticating the future by purportedly giving it a voice – a voice that, given its unaccountability, can only fail to gain genuine legitimacy and authority in the public realm.

Generous futures

The four figures of futurity that this essay has identified as playing an important role in struggles for the temporal commons – foreclosure, obduracy, prefiguration, and future generations – all share a commitment to countering the 'enclosure' of the temporal commons, whereby time is individualised, privatised, and commodified, by practising a politics of generosity towards the future, leaving a positive 'legacy' for those who inhabit the world after us. Such an ethos of generosity is based on a particular approach to the temporality of *justice* as something that cannot be reduced to practices of calculation or equivalence. Nigel Clarke (2013) draws on Georges Bataille's philosophy of solar abundance to articulate forms of ecological justice that are beyond measure, and are based on an excess that animates a sense of justice that is not entrenched within logics of scarcity and calculable harms. Ecological justice, he argues, must be able to respond to events that are so singular that they cannot be encompassed within logics of simple causality or provable culpability.

Such a sense of justice needs to be extended to conceptions of intergenerational justice that concern themselves with justice between living and unborn lives. The widespread arguments over the legitimacy of forms of future 'discounting', for example (where, in cost-benefit policy analyses, future lives are given a lower cost than present lives), reduces the future to a logic of calculation which ignores the extent to which a radically unknowable future contains possible events that exceed any measure. Practising the future as a form of commons requires passing on the future as a gift (Kirwan, 2013), through an ethic of generosity, to *distant* others, not to future lives rendered virtually the same as the present (Derrida, 2006; Barnett, 2005). Intimations of this ethos towards the temporal commons are visible in the figures of solar abundance, persistent struggle, and processual utopianism that offer struggles for the temporal commons' potent sources of power.

In Walt Whitman's 'By Blue Ontario's Shore', he attributes to the poet a talent of judging 'not as the judge judges but as the sun falling around a helpless thing'. The poet judges as falling sunlight. For Jane Bennett, this solar judgement has two qualities. First, it requires magnanimity, an ability to be as accepting of the things the poet encounters as Nature is of him or her. Second, it requires far-sightedness, an ability to see an eternity of fibres stretched out over time: a landscape 'of pulsating threads in a lively field of becoming, always interacting, durational threads' (Bennett, 2011). Practising and protecting the temporal commons, I would argue, demands the invention of forms of intergenerational justice that are based on such solar generosity and far-sightedness.

References

Adkins, L. (2014) 'Speculative Futures in the Time of Debt', paper presented at the conference on *Futures In Question*, Goldsmiths College, University of London, 11–12 September.

Alliance for Future Generations (2015) 'About the Alliance for Future Generations', http://www.fdsd.org/wordpress/wp-content/uploads/abouttheallianceforfuturegenerations signondraftwithoutmembers.pdf, last accessed September 2015.

Amoore, L. (2013) *The Politics of Possibility: Risk and Security Beyond Probability*, Durham, NC: Duke University Press.

Auerbach, E. (2003) *Mimesis: The Representation of Reality in Western Literature*, trans. W. Trask, Princeton: Princeton University Press.

Barnett, C. (2005) 'Ways of Relating: Hospitality and the Acknowledgement of Otherness', *Progress in Human Geography*, 29(1), pp. 5–21.

Bennett, J. (2011) 'The Solar Judgment of Walt Whitman', in *A Political Companion to Walt Whitman*, Lexington, KY: University Press of Kentucky.

Bluedorn, A. & Waller, J. (2006) 'The Stewardship of the Temporal Commons', *Research in Organizational Behavior*, 27, pp. 355–396.

Bresnihan, P. & Byrne, M. (2014) 'Escape into the City: Everyday Practices of Commoning and the Production of Urban Space in Dublin', *Antipode,* 47(1), pp. 36–54.

Brigstocke, J. (2013) 'Immanent Authority and the Performance of Community in Late Nineteenth Century Montmartre', *Journal of Political Power*, 6(1), pp. 107–126

Brigstocke, J. (2014) *The Life of the City Space, Humour, and the Experience of Truth in find-de-siècle Montmartre*, Farnham: Ashgate.

Butler, J. (2011a) 'Bodies in Alliance and the Politics of the Street', *Transversal*, http://www.eipcp.net/transversal/1011/butler/en.

Butler, J. (2011b) 'For and Against Precarity', *Tidal: Occupy Theory, Occupy Strategy*, 1, pp. 2–11, http://tidalmag.org/pdf/tidal1_the-beginning-is-near.pdf.

Chatterton, P., Featherstone, D. & Routledge, P. (2013) 'Articulating Climate Justice in Copenhagen: Antagonism, the Commons, and Solidarity', *Antipode*, 45(3), pp. 602–620.

Clarke, N. (2013) *Inhuman Nature*, London: Sage

Dawney, L. (2013) The Figure of Authority: The Affective Biopolitics of the Mother and the Dying Man, *Journal of Political Power*, 6(1), pp. 29–47

Dean, J. (2012) *The Communist Horizon*, London: Verso.

De Angelis, M. (2007) *The Beginning of History: Value Struggles and Global Capital*, London: Pluto Press.

Deleuze, G. (1992) 'Postscript on the Societies of Control', *October*, 59, pp. 3–7.

Derrida, J. (2006) *Spectres of Marx: The State of Debt, The Work of Mourning and the New International*, trans. P. Kamuf, London: Routledge.

Earth Manifesto (2015) 'A Bill of Rights for Future Generations', http://earthmanifesto.com/1.3%20-%20A%20Bill%20of%20Rights%20for%20Future%20Generations.htm, last accessed February 2015.

Edelman, L. (2004) *No Future: Queer Theory and the Death Drive*, Durham, NC: Duke University Press.

Glennie, P. & Thrift, N. (2009) *Shaping the Day: A History of Timekeeping in England and Wales 1300–1800*, Oxford: Oxford University Press.

Guyer, J. (2007) 'Prophecy and the Near Future: Thoughts on Macroeconomic, Evangelical, and Punctuated Time', *American Ethnologist*, 34(3), pp. 409–421.

Haraway, D. (1997) *Modest_Witness@Second_Millenium FemaleMan_Meets_Oncomouse: Feminism and Technoscience*, New York: Routledge.

Holloway, J. (2010) *Crack Capitalism*, London: Pluto Press.

Ince, A. (2012) 'In the Shell of the Old: Anarchist Geographies of Territorialisation', *Antipode*, 44(5), pp. 1645–1666

Jameson, F. (1982) 'Progress vs Utopia: Or, Can We Imagine the Future?', *Science Fiction Studies*, 27, pp. 147–158.

Lazzarato, M. (2012) *The Making of Indebted Man: Essay on the Neoliberal Condition*, trans. J. Jordan, Cambridge, MA: MIT Press.

Lee, R., Leyshon, A., Aldridge, T., Tooke, J., Williams, C. & Thrift, N. (2004) 'Making Geographies and Histories? Constructing Local Circuits of Value', *Environment and Panning D: Society and Space*, 22, pp. 595–617.

Lejano, R. & Ericson, J. (2005) 'Tragedy of the Temporal Commons: Soil-Bound Lead and the Anachronicity of Risk', *Journal of Environmental Planning and Management*, 48(2), 301–320

Linebaugh, P. (2014) *Stop, Thief! The Commons, Enclosures, and Resistance*, Oakland, CA: PM Press.

Massey, D. (2005) *For Space*, London: Sage.

May, J. & Thrift, N. (eds) (2001) *Timespace: Geographies of Temporality*, London: Routledge.

McKee, Y. (2013) 'Debt: Occupy, Postcontemporary Art, and the Aesthetics of Debt Resistance', *The South Atlantic Quarterly*, 112(4), pp. 704–803.

McKee, Y. (forthcoming) *Strike Art!: Contemporary Art and the Post-Occupy Condition*, London: Verso.

Millner, N. (2013) 'Routing the Camp: Experiential Authority in a Politics of Irregular Migration', *Journal of Political Power*, 6(1), pp. 87–105.

Morton, T. (2013) *Hyperobjects*: *Philosophy and Ecology after the End of the World*, Minneapolis: University of Minnesota Press.

Movement Generation (2011) 'Foreclose on Wall Street', http://movementgeneration. org/join-occupy-sf-foreclose-on-wall-street-west-on-october-12–2011/, last accessed February 2015.

Occupy London (2011) 'Initial Statement', http://occupylondon.org.uk/about-2/, last accessed September 2015.

Occupy Wall Street (2015) 'About', http://occupywallst.org/about/, last accessed February 2015.

Seyfang, G. (2003) 'Growing Cohesive Communities One Favour at a Time: Social Exclusion, Active Citizenship and Time Banks', *International Journal of Urban and Regional Research*, 27(3), pp. 699–706.

Sharma, S. (2014) *In the Meantime: Temporality and Cultural Politics*, Durham, NC: Duke University Press.

Soja, E. (2010) *Seeking Spatial Justice*, Minneapolis: University of Minnesota Press

Springer, S. (2013) 'Human Geography without Hierarchy', *Progress in Human Geography*, 38(3), pp. 402–419.

Stavrides, S. (2013) 'Contested Urban Rhythms: From the Industrial City to the Post-Industrial Urban Archipelago', *The Sociological Review*, 61(S1), pp. 34–50.

Strike Debt (2015) 'You Are Not a Loan', http://strikedebt.org/, last accessed February 2015.

Thompson, E.P. (1969) 'Time, Work-Discipline and Industrial Capitalism', *Past & Present* 38(1), pp. 56–97.

Tomales Bay Institute (2003) *The State of the Commons*, http://www.onthecommons.org/ sites/default/files/stateofthecommons.pdf.

van der Sande (2013) 'The Prefigurative Politics of Tahrir Square – an Alternative Perspective on the 2011 Revolutions', *Res Publica*, 19, pp. 223–239.

Vasudevan, A. (2015) 'The Autonomous City: Towards a Critical Geography of Occupation', *Progress in Human Geography,* 39(3), pp. 316–337.

8 Imaginaries of a global commons

Memories of violence and social justice

Tracey Skillington

Introduction

This chapter examines how imaginaries of 'a global commons' are constructed by the community of the United Nations (UN) on the basis of a witnessing of the violence of atrocity both as it exists today in various regions of the world and as it is remembered from the past. In particular, it will assess how the UN acts as an agent of pedagogic instruction in this regard, exploring the various ways in which particular memories of violence can be seen as 'belonging' to all members of a humanity still plagued by genocidal urges. While the UN's more recent project of remembrance and learning from atrocity is directed in the first instance at co-members of a politically constituted international community, it nevertheless also proves instrumental in initiating a discourse on what makes this community something more than a legal-political entity, something akin to a global commons. The 'unto-itselfness' of human existence (Nancy, 1991: 5) may be driven by the impulse to survive, the desire for autonomy, as well as the realization of personal freedoms, but humankind has always also demonstrated a basic need to make sense of the relationality of our 'being together' (Nancy, 1996) as co-dwellers of one world and explore ideas we share 'in common' (Nancy, 1999: 10), including those as to the unrelenting capacity of people's cruelty to each other. Rarely are memories of barbarism seen as belonging exclusively to particular national or ethnic communities but, rather, they come back to us all (Morin, 2006: 1) as a reminder that vigilance in the constant struggle against genocidal tendencies is the fundamental responsibility of all members of humanity.

Over time, this more underlining common human element (i.e., the desire to establish collective understandings of our co-existence) has contributed to the formation of what Peirce (1998) refers to as the 'commens' or a common mind-set on what are perceived to be the core elements of our social being. Exercises in collective remembrance, like those under investigation here, emerge on the basis of a creative use of this commens and provide a key opportunity for us to explore how certain elements of our communal essence come to be seen as intricately bound up with a propensity to moral evil (Kant, 1998b). Beyond the expectations we have of each other as co-habitants of post-war 'transnational constituencies of peace' (Bohman, 2012), and the normative ideals (international law) that underlie such political and legal alliances (Steger, 2008), are these more fundamental concerns. The narratives in which memories of atrocity are explored typically blend images of horror that are past with counterfactual visions of what is still not the case in the present but what might be, or ought to be, in the future. The normative world of this commons is therefore necessarily multi-temporal. It is compelled to reflect backwards (in memory) and forwards in the hope of finding a meaningful explanation as to why such violence persists in the present (Sarat and Kearns, 2001: 56). What emerges is an understanding of this global commons as situated perpetually within a time frame that is 'convoluted', one whose moral political progress, in being multidirectional, is anything but smooth or linear. 'Learning from the past' comes to be seen as something more than just a registering of the gruesome facts of numerous shameful histories. It also reflects a serious effort to confront humanity's 'worst violence' and use such opportunities to rediscover those other, more positive elements of its 'being-in-common' (Nancy, 1999), that is, its capacities for solidarity, critical reflection, and social transformation.

Learning lessons from atrocity

On 1 November 2005, the UN General Assembly adopted by consensus Resolution 60/7, declaring 27 January as the annual International Day of Commemoration in Memory of the Victims of the Holocaust and requesting the UN Secretary-General to create a new outreach programme to 'mobilize civil society for Holocaust remembrance' and aim to 'instill the memory of the tragedy in future generations' (The Holocaust and the United Nations Outreach Programme, 2013). In the years since, annual commemorations of the Holocaust have explored different dimensions of this tragedy, focusing on themes like that of bravery, rescue, and witnessing, while the newly established UN Outreach Educational Programme has sought to actively extend learning from atrocity beyond the act of remembering, to include also critical reflections on newer threats to peace and the ongoing need for vigilance. Such efforts on the part of the UN reflect 'the layers of a variety of retelling' of one more familiar narrative of human tragedy (Benjamin, 1968: 89), one that begins with the Holocaust but continues on with the Cambodian genocides of the 1970s, those of Bosnia–Herzegovina (1992–94), Rwanda (1994), Darfur (2003–), to connect up with newer scenes of atrocity in Syria and Iraq, all becoming globally recognizable and interchangeable fragments of an ever more common tale of the reoccurring sameness of human barbarism. As a cosmopolitan, transnationally grounded project (Levy and Sznaider, 2006: 23), the UN's Learning Lessons from Atrocity has come, in the years since its establishment, to gather moral insight from many global sources and direct its message to multiple contexts of reception (as opposed to any one particular national scene). Virtual, interactive, and video technologies all facilitate the UN's efforts in this regard to encourage a new degree of global solidarity around certain shared 'epochal commonalities' (Levy and Sznaider, 2002: 93) of genocidal trauma. Digital technologies allow new spaces of communication to emerge around a whole spectrum of barbaric histories that transcend cultures, nationalities, and generations.

What we are offered are twenty-first-century 'illuminations' (Benjamin, 1969) on various episodes that lead from the context of their own morbid history into the analytic and technologically advanced context of a global media present. Uploaded archives of images and oral recollections of their horror lead us back in time to the site of trauma. The sheer 'factuality' and raw empiricism of these representations (Slater, 1995: 223) prove vital to a demonstration of the terror and trauma experienced by the victims of such horror and encourage a greater moral engagement with the persisting violence of atrocity today. As more and more archives are made available for public viewing, the temporal and spatial playing fields of such memories become blurred. Images of Holocaust destruction inspire memories of Bosnia (Zelizer, 1998; Levy and Sznaider, 2002) that, in turn, draw comparisons with Syria (see Vullamy, 2013). The UN's efforts in this regard to provoke a greater dialogue between past and present are greatly aided by the capacities of new technology to allow global audiences to bear witness to the deeply human and banal aspects of multiple atrocities. A critical reconstruction of this dialogue, as it is articulated across different media contexts of communication today, affords us the opportunity to look at how ideas of an intergenerational commons gradually come to be constructed around common experiences of violence, suffering, and human loss. Regret, shame, horror, and fear are not merely states of feeling articulated through media-intense and globally disseminated memory narratives in this instance. Humanity's 'shameful indifference' to and 'abandonment' of millions of victims of atrocities over the years (see Gabriela Shalev, Permanent Representative of Israel to the UN, January 2009) also become the basis of a learning exercise designed to induce a broad consensus that it is 'wrong' that the 'good' do nothing and, in that, allow 'evil to triumph' (see Kofi Annan, General Assembly, January 2005; Navi Pillay, August 2014). The common relational element being referenced here is the freedom of the human will to choose between alternatives. Even as by-standers, we choose to allow barbaric histories to continue. The alternative requires an equal amount of conscious effort. It is to choose to live collectively by a 'law of duty' (Kant, 1998b) that binds all into relationships of mutual respect and common peace.

Central to these learning exercises are annual UN commemorations of the Holocaust and genocide in Rwanda. Memories of both are marked each year with ceremonies, concerts, exhibitions, and video conferences connecting students around the world with representatives of the international Criminal Tribunal for Rwanda, for instance, the National Unity and Reconciliation Commission, as well as genocide survivors. The aim is to initiate a truly transnational exploration of the communal aspects of our 'education through remembrance' (UNESCO, 2013). Screenings of documentaries like *As We Forgive* in UN Centres around the world (16th Commemoration of the Genocide in Rwanda, April, 2010), for instance, are in the interests of promoting a distinctly post-national ownership of memories, one that emphasizes their 'extra-territorial quality' and broader moral significance to a world still scarred by such violence. All the UN's efforts in this regard are focused on a cosmopolitization of memories of Rwanda, the Holocaust, and a number of other atrocities, where the meaning and relevance of these historical events continue to be explored in response to the changing cultural interests as well as social and political circumstances of contemporary global publics. Clearly, the intention is to inspire concerted action and, in particular, greater efforts to prevent future atrocities (see UN Resolution 63/308, 2009). The UN, however, remains burdened with the fact that it does not always deliver on its promise of 'never again'. On the one hand, it yearns for international peace and works consistently towards this realization. On the other, it openly admits that it is limited in its capacities to change a cruel history of repetition.

Arguably, this burdened disposition serves a clear political purpose. Only through a historical perspective that recognizes an unfortunate repetition of violence can the international community attempt to direct moral disenchantment towards a more constructive moment, that is, towards illustrating how the 'now time' of the global present carries with it not only traces of multiple violent pasts but also imaginaries of a reformed future. Because of the sheer intensity of their horror, episodes like the Holocaust, Cambodia, Rwanda, or Bosnia do not pass so readily into a historical 'non-being' but, rather, continue to be explored in terms of their meaningfulness and relevance to the self-identities and cognitions of global communities today. Remembrance comes to initiate a type of 'telescoping' (Benjamin 2002: 471) in upon the details of specific, pasts in the hope that such practices will eventually give rise to a moment of immanent 'world disclosure' (Buck-Morss 1997: 60) when deeper affinities are revealed and a greater clarity is brought to bear on the endurance of violence across time. What continuously destabilizes the perspective of this present is the accumulation of new encounters with catastrophic history (e.g., Syria, Gaza). The violence of genocide returns under particular conditions of collective historical experience as repetition. As Benjamin (quoted in Buck-Morss 1997: 52) comments:

> One used to consider the 'past' [*Gewense*] as the fixed point, and saw the present as attempting to lead knowledge gropingly toward this firm ground. Now the relationship is to be reversed, and the past becomes the dialectical turning, the dawning of the awakened consciousness.

Benjamin's words here may be seen as an important starting point for our discussion on the remembrance practices of the UN and, in particular, its desire to initiate a truly transnational educational project on global histories of atrocity. The working understanding here is that there are certain hidden truths in consecutive histories of atrocity that potentially offer great insight into how a global commons can realize a less violent future.

Understanding temporalities of violence

UN memory narratives typically substitute traditional distinctions between a clear 'yesterday' and an equally well-defined 'today' with the notion of 'lived time', that is, time with a lived present (see Vanhoozer, 1990: 100). In the process, atrocities spread across time and geographical context, including Rwanda (1994),

Cambodia (1975–79), Bosnia (1992–94), 9/11 (2001), or Darfur (2003–) come to be reconfigured as components of a one larger catastrophe that keeps 'piling wreckage upon wreckage' (Benjamin, 2003: 392):

> Despite the pledge so often expressed of 'never again', our warnings were in vain. After the Cambodian massacres, it is Africa that has paid the greatest tribute to the follies of genocide over the past 15 years. After Rwanda, it is Darfur and its dramatic death toll: 200,000 dead and nearly 2 million refugees.
>
> (Simone Veil, 2008: 38)

> The twentieth and the beginning of the twenty-first century has been a time of unparalleled instances of genocide and mass atrocities. The Herero, the Armenians, the Holocaust and other Nazi crimes, Stalinist crimes, Cambodia, East Timor, Rwanda, Bosnia and Herzegovina, Kosovo, Darfur, Democratic Republic of the Congo . . . the list goes on.
>
> (UNESCO, 2013: 20)

What is 'past' and 'present' does not necessarily correspond to the notion of 'before' and 'after' in the episodic structure of this account of the human penchant for violence. Instead, various temporal determinations reverse back into each other (Butler, 1990: 1717) as genocide histories are reordered to create a temporality of violence that is 'convoluted' (Benjamin, 1968) rather than past, one that does not coincide entirely with the notion of cosmological or linear time (Ricoeur, 1984) and, with it, the mythical structure of our progress in time as always moving forward. Alternatively, this representation of violence (and our lack of progress) across territory, time, and historical ruin, can be cognized collectively as a seamless whole. The latter gives form and unity to the identity of this global commons whose repeated encounters with genocide oblige us to rethink the time of moral agency and question whether learning is, in fact, linear and stable. The only constant, it seems, is the rhythm of humanity's downfall:

> The shadow of Srebrenica has joined that of Rwanda, Cambodia, the Holocaust. Each time we hear 'never again'. But do we have the courage to care – and the resolve to act?
>
> (Ban Ki-Moon, 12 January 2013)

A global community in mourning

The ongoing failure of this international community to prevent further atrocities from occurring destabilizes the current temporal regime (i.e. the present) as one after the facts of the worst atrocities. It also inspires profound guilt and a determination to 'never forget' those 'murdered by the enemy' and by 'the world's silence' (Elie Wiesel, UN General Assembly, January 2005). The connectivity of different violent histories is explored chiefly through a historical sensibility that grieves our common and all-too-familiar failure to protect. The dominant image is that of a funeral procession full of sorrow and unfulfilled promises of 'never again', wandering across the globe from one scene of mourning to the next. Just like the heroic yet melancholic figure in the baroque mourning play, today it is the UN who adopts the role of chief melancholic sceptic of humanity's historical disposition and capacity to win the battle against evil's eminent return. In its annually staged commemorative performances, the dead, the living, and the future dead are deliberately suspended in a continuous play of melancholy.[1]

> It is a singular honor to be with you this evening to mark the sixteenth anniversary of the Rwandan genocide, sadly one of the defining events of the 20th century. The macabre and deliberate acts of that April left an enduring stain on human history and on the noble aspirations of this world body. The funeral processions are still winding their way to the Genocide Memorial in Kigali.
>
> (Edward Luck, 16th Commemoration of the
> Rwanda Genocide, 7 April 2010)

> I visited the graves and wept with the mothers of the slain. It is not an easy place for a United Nations Secretary-General to visit. The United Nations – the international community – failed to protect thousands of Bosnian Muslim men and boys from slaughter.
>
> (Ban Ki-Moon, 12 January 2013)

As mourners of multiple, preventable tragedies, we are encouraged to stare into the depths of this all-too-violent world and embrace accumulating numbers of 'lost generations'. Our grief is inspired by a world-historical sorrow that can never fully reconcile itself with 'the disgrace of Auschwitz' (Jonas, 1996), 'the universal shame' of Gaza (Ban Ki-Moon, 31 July 2014, see Sherwood and Balousha, 2014), or the unthinkable butchery of Rwanda. There is no real explanation for the barbarism that continues to grip this world (see Balé, Permanent Representative of the Republic of the Congo to the United Nations and Vice-President of the United Nations General Assembly, January 2013), for the endless 'fabrication of corpses' (Arendt, 1964: 134) and the ongoing debasement of human existence (for example, see Navi Pillay (2014) on the failures of the international community to act in relation to atrocities in Syria). As it ponders such issues, the international community weaves 'the wretchedness of the same in a manner of tragedy' (Benjamin, 1985: 77) into a distinctive narrative of violence across time:

> We have witnessed Cambodia, Rwanda, Bosnia, Darfur, and other genocides – each with its own grim place in the annals of human infamy – but nothing quite the same as the Holocaust's unique reach, its systematized spite, its murderous bureaucracy, its premeditated, purposeful, and planned malice.
>
> (Susan Rice, US Permanent Representative
> to the United Nations, January 2010)

The 'mournfulness' inspired by such reflections on humanity's endless histories of violence evokes a sorrow that points simultaneously downwards into 'the depths of the earth' (Benjamin, 1985: 152), but also upwards to higher intentions, especially moral-ethical duties that we share in common to protect a vulnerable humanity from a similar fate. The fact that 'great tragedies' keep occurring means that our duties to the dead are unresolved and phases of mourning remain incomplete (Derrida, 2001). Collective sentiments of loss, as a consequence, are locked into an indefinite time. Peace may be the goal, but never the full outcome. Not only do we mourn what we have lost in common, but also that which has never been fully achieved, that is, post World War II promises of solidarity, peace, and tolerance. Our anguish is heightened by the realization that genocide victims are global humanity's 'stolen generations', those whose life and contribution to humanity are 'incomplete', and for whose absence we are responsible (through repeated failures to protect). We remain duty bound to preserve the moral presence of these victims as humanity's 'disappeared', those who 'still haunt our collective conscience' (Ban Ki-Moon, 2009) and, in that, are never entirely absent (Derrida, 2001: 44):

> While the lives of so many were taken, their spirit lives on – their dreams have never died.
>
> (H.E. Ron Prosor, Permanent Representative of Israel to the United Nations, January 2013)

> The resounding voices of survivors touch us in ways that no other words could. Yet the silence of the more than 800,000 innocent victims still haunts our collective conscience.
>
> (United Nations Secretary-General Message, 15th Commemoration of the Genocide in Rwanda, April 2009)

Transnationally staged scenes of repentance and mourning become instrumental in initiating a broad discourse of culpability where the international community publicly accuses itself all at once of allowing atrocities to continue. The robbery of 'millions of human souls' has to be understood in some fundamental way as unforgiveable (see Kofi Annan, Former Secretary-General, January 2005). Ultimately, it is this realization that prevents any final reconciliation (or *Versöhnung*) with the murder of innocents. Mourning is focused on the injustice of victims' passing. The latter is compensated (but never excused) by shared efforts to preserve these people, in memory and speech, within the realm of lived time. Above all, victims must continue to be remembered as 'missing' members of the 'global human family' (H.E. Mr. Ron Prosor, Permanent Representative of Israel to the United Nations, January 2013), those who were mercilessly 'torn' or 'ripped away' from the arms of humanity (see, for instance, Ban Ki-Moon, Annual Message on International Day of Commemoration of the Holocaust, 27 January 2012). Such representations heighten the guilty sentiments of a 'universal conscience'

(Derrida, 2001: 33) that stretches out beyond the traditional boundaries of linear time and embraces all generations and nationalities similarly burdened by genocidal histories.

The counterfactual imagination and the quest for redemption

What is also distinctive about UN memory performances is the manner in which they self-consciously point beyond the mere act of remembrance to embrace also a reflection on newer threats to international peace. In the process, the funeral procession becomes an allegory for a wider journey of discovery for humanity at large. The truth of the violence of atrocity is thought to disclose itself in these moments of 'awakening' when newer episodes of violence trigger a reversal in the collective experience of 'time as catastrophe' into a moment of hope for future social transformation. In this moment, allegories of death and ruination suddenly change direction and regroup as a quest for redemption.

> Ultimately, the intention does not faithfully rest in the gaze at bones but leaps faithlessly to the resurrection.
>
> (Benjamin, 1977: 232–233)

The spectacle of a barbaric universe jumps forward to the promise of 'a better future' (The Holocaust and the United Nations Outreach Programme, November 2005). The image of history as a 'hill of skulls' is now supplemented with a desire to 'resurrect the spirit' (Buck-Morss, 1991: 229) and reiterate commitments to justice, as well as a determination to pursue the perpetrators of atrocity. The normative commitments of this commons are always present even though they are also compelled to reflect backwards and forwards, constituted as they are by a tension between memories of the past, present realities, and imaginaries of a different future.

> Suspected *genocidaires* and other would-be criminals around the world now know that they will be held accountable before the International Criminal Court, other international tribunals or domestic courts. The International Criminal Tribunal for Rwanda continues to deliver justice, with the cooperation of Rwanda and other states. International criminal justice is a testament to our collective determination to confront the most heinous crimes. The new age of accountability is real.
>
> (Nineteenth Commemoration of the Genocide in Rwanda, Secretary-General's Message, 2013)

In 2013 commemorations of International Day of Remembrance of the Holocaust, violence in Syria became the focus of delegates' speech:

It is essential that all perpetrators of international crimes understand that they will be held to account. There will be no amnesties for those most responsible. The old era of impunity is ending. In its place, slowly but surely, we are building a new age of accountability. But the important thing is to end the violence in Syria – now – and begin the process of transition.

(Ban Ki-Moon, 12 January 2013)

Such representations of Syria deliberately evoke memories of Nuremberg, Rwanda, and, indeed, Tokyo as 'thick time' is conjured away and new evidence of heinous crimes is presented. The temporal framework created to understand the constancy of violence across time adds 'unity in the synthesis of what is manifold in appearance' (Kant, 1998a: 111) and further reinforces a transcendental quality in shared reflections upon the reoccurring problem of genocide. In such moments of 'reflexive abstraction' (Piaget, 1980: 296; Miller, 2002: 9), the international community stands back from the immediacy of singular episodes of violence and re-affirms the existing order of international legal justice but simultaneously negates half-hearted attempts to realize it. Apart from memories of tragedy, what also unite members of this global commons are shared imaginaries of peace and social justice. Dreams of a more peaceful human existence are not just 'prophetic visions' (Benjamin, 1996) of a world free of 'radical evil' (Kant, 1998b). They are also a work of 'memory justice' (Booth, 2001), and a compulsory one at that. The 'messianic power' (Benjamin, 2003) of such utopian visions is reflected in their capacity both to make sense of the catastrophe of multiple histories of atrocity during and since the Holocaust and also to somehow redeem human history by recasting its tragedies as part of a teleological journey culminating in the establishment of a more just social order (e.g., international law, human rights legislation, etc.). Atrocities continue to inspire the UN's plan for global peace (see Rice, Permanent US Ambassador to the UN, January 2010), even as persisting violence also inspires a certain crisis of confidence in our capacities to demonstrate the political will and 'courage' needed to stop the bloodshed (see 2013 Observance of UN Commemoration in memory of the victims of the Holocaust).

We must find the will and the means to break the cycle of reoccurring violence, hatred and retribution that still afflict too many parts of the world.

(Luck, 16th Commemoration of the Rwanda
Genocide, 7 April 2010)

The accumulating debris of war and genocide teaches us something about the fragility of an international legal order unable to prevent the murder of innocent people. International delegates use remembrance occasions to question why the UN cannot move beyond the ruins of war and realize an alternative history of peace. Frustration is expressed at the lack of consensus on the extent of ethico-political bonds to distant others. What becomes clear, however, is that some form of learning does emerge from these reflections on our collective failure to fulfil duties to protect. Newer commitments to peace are both within (in memory) and external to humanity's histories of atrocity. Without memories of grotesque chapters of violence and terror, the UN's contemporary 'legacy of hope' will not resonate with the sensibilities of the present age. It must promote a message of hope for the future and steer a clear ethical path through bloodied histories by reconnecting a disenchanted world with a human rights-based project of agency, hope, and redemption. As it does so, temporalities of past, present, and future regain their lost connectedness to become elements of a more coherent historical continuum:

> We look back today in shared grief, but we can gather strength in our universal desire to promote peace, tolerance and human dignity ... By remembering the victims of the Holocaust and by recognizing our responsibility to acknowledge and act upon all bigotry and hatred wherever it occurs, we are honouring the values and goals of the United Nations. As members of this family of nations, it remains our collective responsibility to reaffirm the universal message of peace and tolerance to all of humankind.
>
> (Raymond Serge Balé, Permanent Representative of the
> Republic of the Congo to the United Nations,
> and Vice-President of the United Nations
> General Assembly, January 2013)

The allegorical imagination brought to life through this dialectic exchange carries with it a force with transformative potential:

> In a world where extremist acts of violence and hatred capture the headlines on an almost daily basis, we must remain ever vigilant. Let us all have the courage to care, so we can build a safer, better world today.
>
> (Ban Ki-Moon, Memorial Message for
> Holocaust Victims, January 2013)

Incompatibilities between the intended purpose of universal law (global peace) and prevailing realities in Syria, South Sudan, Gaza, the eastern Democratic Republic of the Congo and more demand justification. It is in the construction of 'reasons for' or explanations for the unevenness or deficiencies in human rights application that the developmental historical experiences of the United Nations, as a collective learning subject, are constituted. Clearly, a greater actualization of the demands of democratic reason will not be realized through rhetoric and annual commemorative performances alone. Action in the name of moral duty is also required (Apel, 1997: 96) to initiate the discontinuity of humanity's history as catastrophe. This point is made most forcibly by international delegates like Raymond Serge Balé, Permanent Representative of the Republic of the Congo to the United Nations and Vice-President of the United Nations General Assembly (January 2013), who points to the exemplary importance of the 'unsung heroes of the Holocaust' and other genocides like Cambodia or Rwanda who 'rose above evil to save others', those who demonstrate 'courage, compassion, moral leadership, self-sacrifice, social responsibility, integrity, and righteousness' (UN General Assembly, 25 January 2013). For Balé and, indeed, many other UN officials, the challenge is 'to brush history against the grain' (Benjamin, thesis VII, 2003) and focus on aspects of the past that move the present into a more critical state of interrogation. If we are to meet the project of cosmopolitan peace half way, an alternative political praxis is required, one that actively discontinues cycles of violence and proves genocide to be the historical contradiction of a more peaceful present. Without both a critical interrogation of historical time and a commitment to action, remembrance practices become fatally detached from exercises in critical reflexive learning and human rights protection:

> We do not choose the circumstances in which we live, but we do choose the way we respond to them. We choose. We all choose. Even in the face of the most terrible tyranny, we choose. That is the basis of moral agency, choices large and small that add up to reveal our character. As President Obama put it in Oslo, we must face the world as it is – a world in which human beings can rise to the most astonishing heroism or sink to the most awful depravity – a world in which we must do more than just bear witness – a world in which choices matter.
>
> (Susan Rice, January 2010)

The alternative political praxis being explored here makes certain demands on the global commons. Crucially, it requires a new level of agency (Touraine, 1988) or a willingness on the part of all its members to actively shape the conditions of their co-existence both now and into the future. What ought to bind members into such a project of civic solidarity (Brunkhorst, 2005) is a willingness to make sacrifices for the sake of a more universal system of self-assertion, one that delivers upon promises of human rights of all persons and, in that, manages to exert a greater level of control over our common destiny.

Hypostatized memories of violence: when the meaning of atrocity becomes myth

Looking across the various and honourable efforts of the UN, there is no doubt that learning from past horror is always present as both an aspiration and a reminder of the need to bring a better world to life. Yet memory justice proves again and again to be 'the experience that we are not able to [fully] experience' (Derrida, 1990: 919). The repeated bombing of civilians in Gaza, or the murder and mass displacement of Iraqi Christians, are a painful reminder of this fact, of the way our capacities to learn from the past remain locked within temporal horizons that bind the irrecoverable loss of those who are 'no longer' and the 'not yet' status of peace into endless cycles of repetition (Butler, 1990: 1715). 'Never again!' is promised nervously as a gesture of generosity and openness to a future of peaceful alterity that few believe can be delivered in any kind of sustained manner under the current international political regime. Memories of past atrocities are unable to act as effective conduits of global peace if they remain subject to interpretive trajectories that are insular and fearful of difference. Could it be that the current potential to realize this alternative future that the UN speaks of and tame the human appetite for violence is hampered by the normative insufficiency of hypostatized memory narratives? It is clear that transnational communities of remembrance must remain communities of living interpretation and justification. When a democratic exploration of memory as meaning is replaced with a social and political organization of memory as power, collective capacities to learn are diminished. Memory as meaning is designed to exercise a destabilizing influence on the efforts of power elites, to freeze memory and use it to exonerate the violence of others. The meaning-generating capacities of memories of atrocity must always remain greater than the censoring capacities of power, but when that power resorts to force to insist upon its will, it acknowledges the limits of its own abilities to grasp the fuller meaning of 'learning from the past'.

Violence, as a consequence, remains a constant for this community and, in turn, evokes a variety of uncertainties and ambivalences around the ongoing pursuit of justice. Yet at no point does such ambivalence lead the international community to jettison the goal of greater justice. The latter remains 'a star which beckons us to follow' (Cornell, 1990: 1697), to seek a future in which time is reversed and memories of barbarism cease to be our destiny. Such a leap of fate, however, requires sacrifice, a willingness to explore the depths of 'our human experience' (Bataille, 1985: 74) and break down the kind of distinctions between self and other that hypostatized memory communities insist upon. To try to affix an identity to these memory narratives once and for all would be to deny their mixed element or the '*mêlée*' of their character. It would also be to deny the plurality within each 'ipseity' (Nancy, 1996: 178) and the essentially entangled elements of 'our being in common' (Nancy, 1990) as members of one global commons.

Conclusion

In more recent years, the UN has begun to invest seriously in efforts to forge a new level of social engagement with genocide remembrance and education. What makes this venture more than just a historical investigation into our failed moral-political development during and since the Holocaust is the open acknowledgement that these remembrance practices are intended as core components of a more contemporary, globally coordinated effort to 'work off the past' in the hope of 'preventing future acts of genocide' (see, for example, UN General Assembly Resolution 60/7 'Remembrance of the Holocaust', 2005; UNESCO, 2012a; 2012b). The clear intention behind such measures is to confront both historical and contemporary violence for what it is so that its path can be averted in the future through the medium of open public debate and practices of democratic reform (Habermas, 1998: 19). If the international community, however, is to successfully challenge current trends towards non-communication and racial and ethnic prejudice, it cannot afford to lose sight of its efforts to link the empirical constancy of violence and discrimination with live processes of intercultural communication and the imperatives of democratic reason. What is required as a prerequisite is a communication promoting a transcendental pragmatic conception of co-responsibility for both the construction of our common historical situation and its future democratic reform (Apel, 1993: 20). Such a communication cannot be limited to a fatalistic comparing of 'like with like', but must also stretch to other challenges like deepening tensions between the ethnic communities of contemporary Europe and the Middle East.

To inspire a greater degree of collective self-understanding in the future, ideally, the remit of the UN's 'remembrance through education' programme (UN General Assembly Resolution 60/7 'Remembrance of the Holocaust') should also extend somewhat beyond histories of atrocity to include a discourse of our co-existence with various minorities today; how we address the issues of migration and border control, or that of social integration. Remembrance rituals ought to not only tell us what kind of world we want to leave behind but also what kind of commons we wish to build for the future. What kind of 'united' community of peoples is needed? Surely the primary purpose of a transnational outreach programme is to provide answers to our common predicament? How we learn from the past shapes how we address new challenges in the future. It is essential that the learning potential of this proclaimed 'universal' education programme is not stunted by more hypostatized representations of violent histories that fail to alert contemporary publics to the connections between violence in the past and ongoing cultures of violence and ethnic intolerance in the present. Escaping a historical consciousness that is fractured by a fatal degree of blindness to the continuity of violence requires a perspective that can see the type of tensions surfacing today (reflected, for instance, in the over-enthusiasm of some political leaders to declare Europe's multicultural project an 'utter failure') as temporally anchored beyond the immediate scope of the present. With the world's multiculturalism and diversities still marred by a constitutive inability to accept difference or otherness, any efforts on the part of more powerful political players to deny this fact will ultimately prove detrimental to the UN's endeavours to remember 'past crimes with an eye towards preventing them in the future' (The Holocaust and the United Nations Outreach Programme, 2013). The challenge, as Nancy notes, is for us to figure out our peaceful co-existence, our 'being-with' multiplicity in a singular global world and not become 'locked into a melancholic fixation' with certain pasts (and not others) (Huyssen, 1995: 256). Learning initiatives like those sponsored by the UN must contribute to the formation of genuinely open transnational spaces where more than just a negotiation of the facts of atrocity can occur, but also a mutually sustained learning from alterity, one that allows multiple cultural perspectives and layers of representation of entangled pasts to speak knowingly to 'our common future'.

Note

1 Article 51 of the UN Charter states that the UN can use force only in the case of self-defence, thereby lessening the potential impact that the UN can feasibly have on the prevention of conflict. Article 2(7) of the UN Charter prohibits interference on the part of the UN in anything seen as being within the jurisdiction of state matters. Though it may be important that the authority of the UN does not surpass that of nation-states in

more routine decision-making procedures, it is precisely this lack of a legitimate or legal right to intervene in all troubled situations that has provoked a certain crisis of identity for the UN and a questioning of the ongoing viability of its peace-keeping activities (Habermas, 2004).

References

Adorno, Theodor. (2003) 'Education after Auschwitz', in Ralf Tiedemann (ed.) *Can One Live after Auschwitz?*, Stanford, California: Stanford University Press, pp. 19–23.

Alexander, Jeffrey, Eyerman, Ron, Giesen, Bernhard, Smelser, Neil J. and Sztompka Pietr, (2004) *Cultural Trauma and Collective Identity*, California: University of California Press.

Annan, Kofi. (2005) Official Statement of the UN Secretary-General, 'General Assembly marks 60th anniversary of the Liberation of the Nazi death Camps' (24 January). See: http://www.un.org/News/Press/docs/2005/ga10330.doc.htm (accessed 20 March 2013).

Apel, Karl-Otto. (1993) 'How to ground a universalistic ethics of co-responsibility for the effects of collective actions and activities?', *Philosophica*, 52(2), 9–29.

Apel, Karl-Otto. (1997) 'Kant's Towards Perpetual Peace as historical prognosis from the point of view of moral duty', in J. Bohman and M. Lutz-Bachmann (eds) *Perpetual Peace: Essays on Kant's Cosmopolitan Ideal*, Cambridge, MA: MIT Press: 79–110.

Arendt, Hannah. (1994) *Essays in Understanding, 1930–54*, Jerome Kohn (ed.), New York: Harcourt, Brace and Co.

Balé, Raymond Serge, Permanent Representative of the Republic of the Congo to the United Nations, and Vice-President of the United Nations General Assembly. (2013) Statement made on International Day of Commemoration for Victims of the Holocaust. Memorial Ceremony, General Assembly Hall, 25 January. See: http://www.un.org/en/holocaustremembrance/2013/2013_bale.shtml (accessed 28 March 2013).

Ban Ki-Moon, UN Secretary-General. (2009) Message Fifteenth Commemoration of the Genocide in Rwanda (April). See: http://www.un.org/en/preventgenocide/rwanda/commemoration/2009/sgmessage.shtml (accessed 11 April 2013).

Ban Ki-Moon, UN Secretary General. (2012) Annual Message on International Day of Commemoration of the Victims of the Holocaust (27 January). See: http://www.un.org/News/Press/docs/2012/sgsm14070.doc.htm (accessed 20 March 2013).

Ban Ki-Moon, UN Secretary General. (2013) Memorial Message for Holocaust Victims (January): See: http://www.un.org/News/Press/docs/2013/sgsm14783.doc.htm (accessed 16 March 2013).

Ban Ki-Moon, UN Secretary-General. (2013) Message Nineteenth Commemoration of the Genocide in Rwanda (April). See: http://www.un.org/en/preventgenocide/rwanda/commemoration/2013/sgmessage.shtml (accessed 11 April 2013).

Bataille, Georges. (1985) *Visions of Excess: Selected Writings, 1927–1939*, Minneapolis: University of Minnesota Press.

Benjamin, Walter. (1968) 'The storyteller: reflections on the works of Nikolai Leskov', *Illuminations*, Hannah Arendt (ed.), New York: Harcourt Brace Jovanovich.

Benjamin, Walter. (1985) *The Origins of German Tragic Drama*, John Osbourne (trans.), London: Verso.

Benjamin, Walter. (1996) 'Imagination', in *Walter Benjamin: Selected Writings, 1913–1926. Vol. 1*, Cambridge, MA: Harvard University Press.

Benjamin, Walter. (2002) *The Arcades Project*, Cambridge, MA: The Belknap Press of Harvard University Press.

Benjamin, Walter. (2003) 'On the concept of history', in *Selected Writings, Vol. 4*, Howard Eiland and Michael W. Jennings (eds), Edmund Jephcott et al. (trans.), Cambridge, MA: Harvard University Press.

Bohman, James. (2012) 'Jus post bellum as a deliberative process: transnationalizing peacebuilding', *Irish Journal of Sociology*, 20(1), 10–27.

Booth, James. (2001) 'The unforgotten: memories of justice', *American Political Science Review*, 95(4), 777–791.

Brunkhorst, Hauke. (2005) *Solidarity: From Civic Friendship to a Global Legal Community*, Cambridge, MA: The MIT Press.

Buck-Morss, Susan. (1991) *The Dialectics of Seeing: Walter Benjamin and the Arcades Project*, Cambridge, MA: The MIT Press.

Derrida, Jacques. (2001) *Cosmopolitanism and Forgiveness*, Oxford and New York: Routledge.

Freud, Sigmund. (1957) 'Mourning and melancholia', trans. Joan Riviere, in *The Standard Edition of the Complete Psychological Works of Sigmund Freud*, James Strachey (ed.), London: Hogarth Press and the Institute of Psycho-Analysis. Vol. XIV.

Gilloch, Graeme. (2002) *Walter Benjamin: Critical Constellations*, Cambridge: Polity.

Habermas, Jurgen. (1998) *A Berlin Republic: Writings on Germany*, Cambridge: Polity.

Huyssen, Andreas. (1995) *Twilight Memories: Marking Time in a Culture of Amnesia*, New York and London: Routledge.

Jonas, Hans. (1996) *Mortality and Morality: The Search for Good After Auschwitz*, Evanston, IL: Northwestern University.

Kant, Immanuel. (1998a) *The Critique of Pure Reason: The Critique of Practical Reason and Other Ethical Treatises: The Critique of Judgment*, Paul Guyer and Allen W. Wood (ed. and trans.), Cambridge: Cambridge University Press.

Kant, Immanuel. (1998b) *Kant: Religion within the Boundaries of Mere Reason: And Other Writings*, Cambridge: Cambridge University Press.

Levy, Daniel and Natan Sznaider. (2002) 'Memory unbound: the Holocaust and the formation of cosmopolitan memory', *European Journal of Social Theory*, 5(1), 87–106.

Levy, Daniel and Sznaider, Natan. (2006) *The Holocaust and Memory in the Global Age*, Philadelphia: Temple University Press.

Luck, Edward, Special Advisor on the Responsibility to Protect. (2010) 16th Commemoration of the Rwanda Genocide, 7 April. See: http://www.responsibilitytoprotect.org/index.php/component/content/article/35-r2pcs-topics/2757-commemoration-of-the-rwandan-genocide-global-centre-brief-on-rtop-and-nigeria#Genocide (accessed 20 March 2013).

McGettigan, Andrew. 2009. 'As flowers turn towards the sun: Walter Benjamin's Bergsonian image of the past', *Radical Philosophy*, 15 (November/December), 25–35.

Miller, Max. (2002) 'Some theoretical aspects of systemic learning', Institut für Soziologie Universität Hamburg Allende-Platz-1 D-20146 Hamburg https://www. miller. EML/1_multipart_xF8FF_1_systemic_learning.pdf/C58EA28C-18C0-4a97-9AF2-036E93DDAFB3/systemic_learning.pdf? (accessed 4 June 2012).

Mitreva, Ilinka, Minister for Foreign Affairs of the former Yugoslav Republic of Macedonia, UN General Assembly. (2005) UN Sixtieth Anniversary of the Liberation of the Nazi Death Camps (January) http://www.un.org/News/Press/docs/2005/ga10330.doc.htm (accessed 28 May 2012).

Morin, Marie-Eve. (2006) 'Putting community under erasure: Derrida and Nancy on the plurality of singularities', *Culture Machine*, 8. Open Humanities Press, pp. 1–9.

Nancy, Jean-Luc. (1991) 'Of being-in common', in *Community at Loose Ends: Miami Theory Collective*, Minneapolis: University of Minnesota Press, 1-12.

Nancy, Jean-Luc. (1993) *Le Sens du Monde*, Paris: Galilée.

Nancy, Jean Luc. (1996) *Être singulier pluriel*, Paris: Galilée.

Office of the High Commissioner for Human Rights. (2014) 'Media Centre – Pillay castigates "paralysis" on Syria as new UN study indicates over 191,000 people killed' (August 22, 2014). See: http://www.unog.ch/unog/website/news_media.nsf/(httpNews-ByYear_en)/6F3D0ADEFB81D8F0C1257D3C002DAE30?OpenDocument#sthash.ucGjqGHO.dpuf (accessed 24 August 2014).

Oskanian, Vartan, Minister for Foreign Affairs of Armenia, UN General Assembly. (2005) UN Sixtieth Anniversary of the Liberation of the Nazi Death Camps, (January) http://www.un.org/News/Press/docs/2005/ga10330.doc.htm (accessed 28 May 2012).

Peirce, Charles S. (1998) *The Essential Peirce: Selected Philosophical Writings. Volume 2 (1893–1913)*, EP 2, edited by The Peirce Edition Project, Bloomington: Indiana University Press.

Piaget, Jean. (1980) *Experiments in Contradiction,* Chicago: Chicago University Press.

Prosor, Ron, Permanent Representative of Israel to the United Nations (2013). International Day of Commemoration for the Victims of the Holocaust Memorial Ceremony, General Assembly Hall (25 January). See: http://www.un.org/en/holocaustremembrance/2013/2013_prosor.shtml (accessed 20 March 2013).

Rice, Susan E., US Permanent Representative to the United Nations. (2010) 'From memory to resolve', International Day of Commemoration in Memory of the Victims of the Holocaust, at Park East Synagogue, New York City, 23 January. See: http://usun.state.gov/briefing/statements/2010/135897.htm (accessed 20 March 2013).

Ricoeur, Paul. (1990) *Time and Narrative. Volume Three*, Chicago: University of Chicago Press.

Ricoeur, Paul. (2003) 'Memory, history, forgiveness: a dialogue between Paul Ricoeur and Sorin Antohi'. See: http://www.janushead.org/8-1/ricoeur.pdf (accessed 20 March 2013).

Sarat, Austin and Kearns, Thomas. (2001) 'Making peace with violence: Robert Cover on law and legal theory', in A. Sarat (ed.) *Law, Violence and the Possibility of Justice*, Princeton: Princeton University Press, pp. 49–84.

Sherwood, Harriet and Balousha, Hazem. (2014) 'UN: "This is a source of universal shame. Today the world stands disgraced"', *Irish Times* (31 July). See: http://www.irishtimes.com/news/world/middle-east/un-this-is-a-source-of-universal-shame-today-the-world-stands-disgraced-1.1883153 (accessed 22 August 2014).

Slater, Don. (1995) 'Photography and modern vision', in Chris Jencks (ed.) *Visual Culture*, London and New York: Routledge.

Taylor, Charles. (2004) *Modern Social Imaginaries,* Durham and London: Duke University Press.

The Holocaust and the United Nations Outreach Programme. See: https://www.un.org/en/holocaustremembrance/ (accessed 15 March 2013).

Touraine, Alain (1988) *Return of the Actor: Social Theory in Postindustrial Society*, Minnesota: University of Minnesota Press.

UN General Assembly. (2005) Resolution 60/7 'Remembrance of the Holocaust' (1 November). See: http://www.un.org/en/holocaustremembrance/docs/res607.shtml (accessed 30 March 2013).

UNESCO. (2012) 'The international dimensions of Holocaust education', International Day of Commemoration in Memory of the Victims of the Holocaust, Conference (31 January). See:

http://www.unesco.org/new/fileadmin/MULTIMEDIA/HQ/ED/pdf/Programme-holocaust-enfr.pdf (accessed 20 March 2013).

UNESCO. (2013) 'Why teach about the Holocaust?' (23 January). See: http://unesdoc.unesco.org/images/0021/002186/218631E.pdf (accessed 20 March 2013).

UNESCO and the Topography of Terror Foundation. (2012) 'The globalization of Holocaust education' (27 April). See: http://www.unesco.org/new/en/education/resources/online-materials/single-view/news/the_globalization_of_holocaust_education/(accessed 20 March 2013).

UNESCO Education for Holocaust Remembrance (2013). See: http://www.unesco.org/new/en/education/themes/leading-the-international-agenda/human-rights-education/holocaust-remembrance/ (accessed 13 April 2013).

USC Shoah Foundation. (2013) The Institute for Visual History and Education. See: http://sfi.usc.edu/education/iwitness/ (accessed 13 April 2013).

Vanhoozer, Kevin, J. (1990) *Biblical Narrative in the Philosophy of Paul Ricoeur: A Study in Hermeneutics and Theology*, Cambridge: Cambridge University Press.

Veil, Simone. (2008) 'The Shoah: a survivor's memory – the world's legacy'. See: http://www.un.org/en/holocaustremembrance/The_Shoah.pdf (accessed 15 March 2013).

Vullamy, Edward. (2013) 'I've seen the horrific result of western paralysis. It musn't happen in Syria', *Observer* (17 March 2013).

Weisel, Elie (2005) General Assembly Marks 60th Anniversary of Liberation of Nazi Death Camps UN General Assembly. (January). See: http://www.un.org/News/Press/docs/2005/ga10330.doc.htm.

Zelizer, Barbie. (1998) *Remembering to Forget: Holocaust Memory through The Camera's Eye,* Chicago: University of Chicago Press.

Part IV

The capture of the commons

9   The matter of spirituality and the commons

Claire Blencowe

A poem by Gloria Anzaldúa[1]
A Sea of Cabbages
(for those who have worked in the fields)

On his knees, hands swollen
sweat flowering on his face
his gaze on the high paths
the words in his head twinning cords
tossing them up to catch that bird of the heights.
Century after century swimming

with arthritic arms, back and forth
circling, going around and around
a worm in the green sea
life shaken by the wind
swinging in a mucilage of hope
caught in the net along with la paloma.

At noon on the edge
of the hives of cabbage
in the fields of a ranchito in Tejas
he takes out his chile wrapped in tortillas
drinks water made hot soup by the sun.
Sometimes he curses

his luck, the land, the sun.
His eyes: unquiet birds
flying over the high paths
searching for that white dove
and her nest.

Man in a green sea.
His inheritance: thick stained hand
rooting in the earth.

His hands tore cabbages from their nests,
ripping the ribbed leaves covering tenderer leaves
encasing leaves yet more pale.
Though bent over, he lived face up,

the veins in his eyes
catching the white plumes in the sky.

Century after century flailing,
unleafing himself in a sea of cabbages.
Dizzied
body sustained by the lash of the sun.
In his hands the cabbages contort like fish.
Thickened tongue swallowing

the stench.

The sun, a heavy rock on his back,
cracks,
the earth shudders, slams his face
spume froths from his mouth spilling over
eyes opened, face up, searching searching.

The whites of his eyes congeal.
He hears the wind sweeping the broken shards
then the sound of feathers surging up his throat.
He cannot escape his own snare –
faith: dove made flesh.

Introduction: the matter of spirituality

Spirituality is the movement of a soul beyond the boundaries of its own identity, the movement of perception beyond the perceptive capacities – the worlded realities – of the perceiver. It is the recognition of the existence of somethings radically other, the sure knowledge of unknowability. Spirituality decentres the self; it is calling to think, feel and act interestedness in others. As active principle it intervenes, distributes, partitions; it sets individuated things – vegetable, animal, mineral, technical – *in relation*. If we have a relationship with land we have a spirituality, or we have at least a 'foyer' – a framework, hearth and home – for spiritual experience. Bent over the earth we live face up, catching glimpses of the spirit's creative movement. Concentrating on a natural form – cabbage, rock, soil – we cannot but confront complexity, mystery, an infinity of layers. As our back is scorched or our fingers are frozen, and our bones ache with the weight of moving or traversing the ground, we know, really *know*, that we are not all there is, that we are limited creatures, that we exist in and by the grace of a cosmos: the dove is made flesh.

Commons are profoundly spiritual. They are sites saturated with significance. Partaking in a commons means moving in currents of transcendence, seeing in relation, acting outside of the self. The spiritual nature of commons is most obviously manifest in the common farming of common land – where holy ghosts and other feathered creatures spring so readily from the soil. But all organisations of objects in common harbour a spiritual capacity. And spirituality helps to harbour the capacity of becoming or persisting as commons. The spirituality of the commons energises commons movements, resonating with religiosity. It binds bodies to a form of life and infuses everyday political practices of commoning with meaning. Spirituality is a snare, a lure to investment, affective grip – twinning rough fibres of feeling into cords that secure frame-works or heave virtual futures into life.

This spirituality is a crucial aspect of the appeal of the commons in our alienated age. Commons promise escape from the disenchanted iron traps of instrumental-ised, privatised lives – a route to self-transcendence that is also the transcendence of nihilism, existential angst and hopelessness. Alienation and disenchantment are not, as is often thought, the symptom of an overly institutionalised or fixed formation of life. Rather, they express the absence of a shared reality, of knowing together in a common world. They are the condition of 'worldlessness' (Arendt 1958), the absence of objectivity (Read 2011, Blencowe, Brigstocke and Noorani 2015). The commons offers precisely a world in common, a public thing (Honig 2012), an outside of subjectivity through which perception, action and attention come to matter. The tragedy of enclosure is the poverty of the soul.

The mattering of commons has an enhanced appeal for our supposedly 'post-secular' moment, in which thirst for politicised religiosity is recognised but widely associated with aggressive identity politics and the 'passionate intensity' of 'the worst'.[2] The investment of lives with matter (significance) through mat-ter (objects, spaces, technologies) held in common promises formations of spir-ituality that are consistent with liberal commitments to openness, inclusion and equality. Instead of transcending the self by becoming a part of the mega-self of a particular identity group or community, the commons offer material practical sites through which we get outside of ourselves by entering, and sharing, a world. To have some-*thing* in common supplants the need to *be* common.

But spirituality *is a snare.* We can choke on our faith. The spirituality of the commons might tempt a naïve imagining that commoning is somehow immune from corruption, exploitation or entrapment. Whether it is communitarian or com-munist, totalising or spacing, spirituality generates energy that can be captured and utilised – and so captured, territorialised, spirituality becomes a means of capture.[3]

The spirituality of the commons

> Commoning has always had a spiritual significance expressed as sharing a meal or a drink, in archaic uses derived from monastic practices, in recogni-tion of the sacred habitus. Theophany, or the appearance of the divine princi-ple, is apprehended in the physical world and its creatures. In North America ('Turtle Island') this principle is maintained by indigenous people.
>
> Peter Linebaugh (2014: 13–14)

The commons and commoning practice are sources of spirituality. This spirituality is not simply a projection of human imagining – it is an expression and efferves-cence of the material and relational constitution of bodies and their forces, in which people are drawn outside of themselves and held in active relations. It is the experi-ence of material interdependence, in which people might be dependent on objects, creatures, land or ecologies as much as on other people. In turn, the spirituality of the commons is crucial in producing and maintaining commons. The spirituality of the commons inspires people to act and it helps to sustain the great energy that is always required in the ongoing production of common space and resource. Faith enables people to perform miracles, to raise new worlds from scorched stretches of earth, to build love and community from hatred or indifference (Katongole 2011). We might think of commoning as so many multiple miracles.

The spirituality of the commons, and the role of spirituality in maintaining a common, is most obviously evident where it intersects with explicitly religious discourse – as in the examples alluded to by Peter Linebaugh above. In this volume Naomi Millner (Chapter 2) explores the aesthetic and, we might say, spiritual attitudes that shaped and fuelled the Open Spaces Movement of the nineteenth century – which she controversially claims for a diverse and disputed history of commons movements. Alongside the mythic Romanticism associated with John Ruskin, the non-conformist Christian sensibilities and theological perception of activists like Octavia Hill were crucial to the inspiration and energy of the movement – a movement with profound legacies of parkland and the UK National Trust.

In the spring of 2012 I attended a symposium (somewhat ironically hosted by the Warwick Business School) which was a platform and discussion space for the London Occupy Movement Outreach Team, which had just been initiated. In discussion with the team members, all of whom had been key figures in establishing and maintaining the occupation of St Paul's Cathedral and other common spaces in London, I was struck by the passionate friendships and mystical intensity binding the diverse group. Love was central to their account both of their experience and their ambition, a discourse powerfully resonant with descriptions of mystic communion. Sharing food and living space was, they maintained, crucial to this new form of being together that they had come to engender. The team talked freely about the spirituality of their experiences in the movement, citing this as a core aspect of their strength and distinction. As they saw it, this open spirituality enabled them to cross the atheist/religious divide, bringing together deeply religious activists and organisations with Marxists and anarchists – groups that have conventionally seen themselves as deeply opposed.

But the spirituality of the commons is not only expressed through explicitly religious discourse. Patrick Bresnihan and Michael Byrne (2015) describe the practices involved in the creation and continuation of independent spaces – urban commons – in contemporary Dublin. Independent spaces require, they maintain, an alternative subjective relationship to space that operates outside of norms of private property. The independent spaces are not constituted once and for all at their inception, and nor are they constituted solely through arrangements of co-ownership. Rather, they require perpetual work of commoning – acts of non-monetary exchange, gifting and sharing. Vast amounts of voluntary labour go into sustaining any urban commons (Bresnihan and Byrne 2015:10). In this volume Bresnihan (Chapter 4) elaborates further on the nature of commoning practices and the relationships they engender, drawing here on his ethnographic researches amongst fishermen. He argues that these fishermen, in their vulnerability and interdependence, constitute commons both at sea and on shore – a commons that is constituted through commoning. The lives of the fishermen were characterised by perpetual 'small acts of generosity':

While they were often incidental and taken for granted, they constituted an invisible network of favours and gifts that operated like a reserve to be drawn on at any moment of need or crisis. Nor were these gifts, of time, labour, resources, calculated in the terms of straightforward utilitarianism; they were not the actions of individuals working out exactly what was in his or her own self-interest. They were the actions of people who knew intimately and immediately that they were part of a wider collective on which they relied.

(Bresnihan, Chapter 4, this volume).

Whilst he does not use the term, the 'spirituality' of the commons is captured well in this sense of being part of the meaning of the relationship. To use more theological language, we could describe these scenes as suffused with grace, charisma and love.

However, religion does figure explicitly in the examples that form the centre of this chapter – the early commons movements of the Diggers, Levellers and Pirates as described in Peter Linebaugh and Marcus Rediker's famous book *The Many-Headed Hydra: The Hidden History of the Revolutionary Atlantic* (Linebaugh and Rediker 2000).

The monster and mystic communalism

That we may work in righteousness, and lay the foundation of making the earth a common treasure for all, both rich and poor. That every one that is born in the land may be fed by the earth, his mother that brought him forth, according to the reason that rules in the creation, not enclosing any part into any particular land, but all as one man working together, and feeding together as sons of one father, members of one family; not one lording over another, but all looking upon each other as equals in creation. So that our Maker may be glorified in the work of his own hands, and that every one may see he is no respecter of persons, but equally loves his whole creation, and hates nothing but the serpent. Which is covetousness.

The 'Digger Manifesto' *The True Leveller's Standard Advanced*
(1649) (cited in Linebaugh and Rediker 2000: 85)

The Many Headed Hydra sets out a new account of the revolutionary forces of seventeenth- to nineteenth-century Britain and the Americas. In contrast to histories that emphasise the agency of the bourgeoisie in the generation of the ideas and energies that fuelled the 'Age of Revolutions' and the foundation of modern political structures, Linebaugh and Rediker set out a classic 'people's history' argument, emphasising bottom-up transformation and the agency of the proletariat in the making of modernity. However, contra those 'people's histories' that identify 'the proletariat' with the white working class, or assume that folk history must pertain to the folk of a particular territory, Linebaugh and Rediker emphasise the transatlantic, transnational, character of the proletariat. They map out concrete links and the movement of ideas between workers' revolts and uprisings across all edges of the Atlantic, citing sailors as 'a vector of revolution' carrying ideas, resources and agency (Linebaugh and Rediker 2000: 241). Through their narrative Linebaugh and Rediker seek to reverse the divide-and-rule tactics through which the interests of the (supposedly white) proletariat in history have been pitted against those of slaves and black people – reverberating the insistence of the seventeenth-century Diggers in England and the nineteenth-century revolutionary writings and preachings of Jamaican Robert Wedderburn that the only real political choice is that between the commons and slavery.

In the seventeenth and eighteenth centuries the image of the many-headed Hydra was often used in the writings of the powerful to characterise the emergent proletariat. The Hydra is a mythical monster that Hercules had to fight as one of his great tasks. When the head of the Hydra is chopped off it grows two new heads in its place. Workers' revolts were depicted as the heads of the Hydra that keep popping up, however often they are put down. By using this imagery the bourgeoisie was defining itself as 'the hero', engaged in a Herculean battle to subdue the monster and establish a new world order. In 1795 Thomas Malthus described the displaced people (lost commoners) of England as akin to the barbarians, the 'hydra-headed monster' that had invaded and destroyed Rome (Malthus 1795: 276). We can see here the image of a 'many headed' body of the monster feeding into the modern concept of the population 'the body with so many heads they cannot be counted' (Foucault 2007).

Linebaugh and Rediker (2000) aim to reclaim this image of the many-headed Hydra from its monstrosity, setting out an affirmative vision of a many-headed transnational and irrepressible revolutionary proletariat. This proletariat emerged with the displacements and dispossessions of peoples through enclosures, clearances and the transatlantic slave trade, and pulled itself together in the ships of the merchants and the British navy which piloted the new factory model of disciplined labour. This proletariat was a 'motley crew', a multi-racial mixture of displaced people, both men and women, circulating around the Atlantic. It was revolutionary, resisting the enclosure of common lands in the practices of the Diggers; fuelling regicidal revolution in England and demanding the radically democratic levelling of all people and the abolition of slavery in the Putney Debates; resisting forced labour and other impositions in countless urban insurrections; fleeing the floating factories to form the maritime commons of pirate ships; rising up against slavery across the Americas; overthrowing slavery in the first-ever successful workers' revolution, that of the slaves in Haiti; and fomenting the American Revolution. And this insurrectionary 'motley crew' was, radically, *religious*.

The Diggers and the Levellers of seventeenth-century England were closely associated with radical religious movements of the day, including Anabaptists, Quakers and 'Ranters'. As can be seen in the extract from the manifesto quoted above, the Diggers drew inspiration from scripture and theological insight as well as their own experience. Bibles in the vernacular were still a relatively recent innovation and persisted as a source of radicalism, despite conservative forces of the Counter-Reformation which defined Anabaptists as heretics fit for deportation or destruction. The biblical theme of God's abundance provided a powerful critique of the ideology of scarcity that was being used to justify enclosures and suffused the understanding of the common with that of divine glory. The idea of 'levelling' – making all equal – was crucially informed by the biblical theme of equality, especially the phrase that God 'is no respecter of persons' (Linebaugh and Rediker 2000: 88). Many Anabaptists professed *antinomianism*, which takes up Corinthians 6:12 'All things are lawful unto me', interpreting this as the statement that faith, which is the manifestation of God in individuals' consciousness, is beyond the law. They believed that 'the 'moral law [was] of no use at all to believers', that the Old Testament was not binding on God's chosen, and that faith and conscience took priority over good works and lawfully constituted authority' (*ibid:* 66). This belief in the primacy of conscience over established law was a powerful resource in imagining revolutionary radicalism and fostering resistance to hierarchy. 'Skepticism toward rules, ordinances, and rituals abounded, as did revelations and visions. Some religious radicals asserted that the "body of the common people is the Earthly Sovereign"' (*ibid*). In the transatlantic 'motley crew' this radical Christian religiosity met with African spirit theologies, generating a powerful catalyst for change that resonates today in contemporary black liberation theology and Rastafarian spirituality (Taylor 2004).

Theologian Mark Lewis Taylor associates the religiosity of the Diggers, Levellers and Pirates, as described by Linebaugh and Rediker (2000), with a concept of 'mystic communalism' rooted in Christian understandings of the Holy Spirit – which, he maintains, are compatible with African notions of spirit. Taylor argues that a thorough survey of the biblical and theological writings on the nature of the Holy Spirit leads to an understanding that living in community has a mystical meaning and power. It is when living and acting in community that the creative force of God within the world – the Spirit – is manifest. He argues that the term '"Holy Spirit" refers primarily to the mystery of God as intrinsic to, immanent in, communal life and development' (Taylor 2004: 379), and that in biblical examples the Holy Spirit is integrally bound up with the creation and nurture of communities of agapic love. The mystical meaning of spirit is 'located in the experience of love in a communal ethos . . . It is a mystical practice where transcending experiences of the sacred, paradoxically, spring up most dynamically in ways immanent to concrete human experiences of agapic community' (*ibid*: 380).

Taylor claims that the essential connection between the Holy Spirit and community is expressed in numerous definitions or attempts to name the 'person' of the Holy Spirit. In trying to explain the concept of the Trinity (the threefold nature of God as the Father, Son, and Holy Spirit), St Augustine experimented with an imaginary in which the Father is 'the Lover', the Son is 'the Beloved' and the third person is 'the Love' that is between them. In contemporary theology Paul Tillich frequently discussed the Holy Spirit as '"Spiritual Community," an ideal community realised in history as one of faith and love, under the "biblical picture of Jesus as the Christ"'; whilst Sallie McFague emphasises 'the spirit of God as "basically and radically immanent" and *in relations* of "love and empowerment, of life and liberty, for people and for the natural world"' (Taylor 2004: 382).

In Taylor's description, the mysticism of the Holy Spirit in Christian communities appears as an expression of something very much like the spirituality that is intrinsic to the commons and commoning. A crucial point to draw from his theological exploration is that the mystical experience, the manifestation of divine creativity and the miraculous capacity that constitute the Spirit derive from – or at least become possible within – the concrete experiences of living in common. This suggests that it was not only a case of radical religiosity fuelling the revolutionary fires of the motley monster of the seventeenth-, eighteenth- and nineteenth-century Atlantic, but that it was also a case of the concrete experience of people thrown together, living in common as a practical and political necessity, generating the mystical experiences of Spirit that underpinned religious radicalism.

The holy spirit or the commons out of place

> Now the Spirit spreading itself from East to West, from North to South in sons and daughters is everlasting, and never dies; but is still everlasting, and rising higher and higher in manifesting himself in and to mankind.

> I have writ, I have acted, I have peace: and now I must wait to see the spirit do his own work in the hearts of others, and whether England shall be the first land, or some other, wherein truth shall sit down in triumph.
>
> Gerrard Winstanley 'the most articulate voice of revolution during
> the late 1640s' (Linebaugh and Rediker 2000: 140–142)

Christian theologies vary very considerably in all manner of ways. One of the key divisions concerns the different emphases that are placed on the nature of God. As Dieter Werner explains, there are three different emphases in the understanding of God that characterise the three main prototypes of church traditions in World Christianity:

> The three emphases are: the emphasis on God as proclaimed Word of Jesus Christ (Protestant tradition), the emphasis on God as charismatic power and source of energy through the Spirit (charismatic and Pentecostal tradition) and the emphasis on God as Eucharistic mystery or transformed substance as celebrated in the sacrament(s) (Orthodox and Roman Catholic traditions)
>
> (Werner 2013: 98).

Taylor points to the specifically *Spirit-centred* nature of the theology of the Diggers, Levellers and Pirates. He thus draws attention to resemblance between these theologies and contemporary Pentecostal and Charismatic traditions. Indeed, one of his key points in discussing the religiosity of the movements described by Linebaugh and Rediker is to reclaim the politics of Pentecostal Christianity from being automatically associated with the political Right. He wants to affirm a long tradition of leftist Pentecostalism, of which black liberation theologian Robert Beckford (1998) is exemplary in the present (Taylor 2000: 190).

Spirit-centred theology emphasises the immanent creative force of God within the world, as well as the capacity of people – or those people who are reborn in Christ – to participate in that creative agency (or charisma). Whilst this certainly can and does have individualistic and identitarian manifestations (as figured in the Prosperity Gospel and the Culture Wars), there is also a radical and democratic impulse to Spirit-centred theology. The idea that human creativity is a manifestation of Spirit informs traditions of radical pedagogy and participatory democracy, as represented in the figure of Paulo Freire (1970) – the idea that people do not have to be told what to do, because if they are truly free to engage their creativity, humanity or 'dignity' they will be enacting the divine agency of Spirit. If people are considered as participants in the creative force of God in the world (the Spirit), then the actions of people have that divine self-justifying, sovereign nature. As radical transformation is a sign of Spirit – a manifestation of its divine creativity – Spirit-centred theology can nurture radicalism. In so far as other people's creative agency can be considered participation in the work of Spirit, that agency can be trusted. We can see ourselves as part of a common project – the work of the Spirit – even though we are acting individually and without instruction.

We see something of this democratic impetus in the above quotes from the seventeenth-century English revolutionary and Digger Gerrard Winstanley. Winstanley is attempting to revolutionise the world, and yet he 'has peace', having made his own contribution; he trusts that 'the spirit [will] do his own work in the hearts of others' (cited in Linebaugh and Rediker 2000: 141). The trust that Winstanley is willing to place in the hearts of others is crucial to the radically democratic levelling spirit that he articulates politically. It relies upon the sense that his agency and ambition are bound up with that of others, that he and others are co-participants in a common creative force – the trans-personal and indeed transnational – life of the Spirit which 'is spreading itself East and West from North to South in sons and daughters is everlasting and never dies' (Linebaugh and Rediker 2000: 142).

Taylor claims that there is a correlation between people who are oppressed and seeking liberation and Spirit-centred theologies or the Holy Spirit more specifically. The Spirit 'thrives among peoples in resistance [and] is associated with their roles as shifting, often forcibly displaced, moving from place to place, mixing cultural ways from continent to continent' (Taylor 2004: 389). For Taylor, then, the Diggers, Levellers and proto-Pentecostal Pirates are examples of the general phenomena of the Spirit becoming manifest amongst displaced and resisting people. There is strong resonance between this identification of the prevalence of Spirit-centred theology amidst displaced, nomadic and mobile people and the identification of such peoples with creative 'lines of flight' (Deleuze and Guattari 1988) and the autonomous generation of immaterial commons *as* collective life force of the multitude (Hardt and Negri 2000).

For Taylor, the reason for this correlation between displaced resisting people and the Spirit is to be found in the nature of the Spirit itself. Drawing on Tillich (1967), he claims that the Spirit is intrinsically liberating or freeing – that it is divine life [which is] in itself (and not just for the world), distinguished by that freedom. Hence the divine life that is believed to act in history, etched deeply into the dynamism and structure of all creation, is a veritable puls-ing of freedom, a resource for catalysing change in the present (to varying degrees) or change in an eschatological or apocalyptic future.

(Taylor 2004: 384)

Taylor's explanation for the correlation between displaced peoples and Spirit is difficult to accept for a number of reasons. Even if we can accept such essen-tialist metaphysical reasoning about the nature of Spirit, the assumption that the Spirit is the same as the force of liberation (whilst commonplace and reso-nant with many vitalist philosophies, such as that of Hardt and Negri) is clearly false. This is because of the evident and powerful role of Spirit in so many religious-political movements that patently *oppress* people and lands. In the Lord's Resistance Army we can see the Spirit rising up amidst displaced and alienated people of Northern Uganda who are resisting their circumstance and pulling together, but could we really describe that Spirit as 'a veritable pulsing of freedom'? The same question could be asked of the Prosperity Gospel or the 'God Hates Gays' brigade.

In a more materialist or sociological reading we can flip the explanation around and suggest that the association between the Spirit and displaced people is not about the Spirit bringing, or even constituting, the movement of liberation – but, rather, is something that comes about due to the absence of concrete resources and places in which to manifest the spirituality of the commons. We might think of the Spirit as the manifestation of the spiritu-ality of the commons *in the absence of a physical common* – constituting a plane of immanence, a site of participation, for people who are not in the position to be participating in the concrete active life of land or inde-pendent spaces. Spirit, then, is an effect, not cause, of mobility, transience and alienation – movements that *may* be, but are certainly not necessarily or ordinarily, movements of liberation. The Spirit might be thought as an immaterial materialisation of the common life of persons who do not have the common object of land or place through which to get outside the self. Ripped from their actual common land, we might speculate, the Diggers and Levellers encountered and made manifest the spirituality of that com-mon in the shared, impersonal/trans-personal life of the Holy Spirit, relat-ing to the world and to each other through the impersonal vital mediator of determination that is Spirit. Rather than 'the veritable pulsing of freedom', then, we might rethink the Holy Spirit as the spirituality of the common *out of place*. In this reading the Spirit becomes an answer to the question of how to constitute a common in the absence of concrete common land or resource. The association between the Spirit and mobility is, then, not due to the emancipatory nature of Spirit but, rather, due to the way that Spirit movements arise from the experience of *dispossession*.

The return of the monster

> If you should destroy these vessels, yet our principles you can never extin-
> guish, but they will live for ever and enter into other bodies to live and act
> and speak.
>
> The Quaker Edward Burrough to Charles II in 1660
(cited in Linebaugh and Rediker 2000: 135)

Linebaugh and Rediker claim that 'the truth' and 'everlasting gospel' preached by
Winstanley and the Diggers persisted across centuries in the transatlantic imag-
inary, returning to England in the nineteenth century in the words of Ottobah
Cugoano and William Blake. In the meantime:

> It sat in swampy tri-isolate communities; it swayed on the decks of deep-sea
> ships; it rubbed shoulders with the poor in the taverns of the divaricated port
> cities; it strained for a hearing on the benches of the Great Awakening, or on
> stools on the dirt floors of slave cabins at night.
>
> (Linebaugh and Rediker 2000: 143–144)

The Great Awakening was a charismatic movement of Protestant revival that took
place in America in the early eighteenth century (and then again on a more dramatic
scale at the post-Independence end of the century in the 'Second Awakening'),
which renewed the Protestant affirmation of the primacy of the individual relationship with God over the mediation of institutions or dogma (MacCulloch,
Bancroft, and Salt 2010). The Awakenings gave rise to the plethora of American
Evangelical churches and paved the way for contemporary Pentecostalism and
Charismatic revival. In contrast to the Reformation of Luther and Calvin two centuries earlier, the Great Awakening emphasised an *emotional* relationship with
God – communing with the movement of the Holy Spirit in the intensity of transformative and spiritual experience. Charismatic preachers excited radical emotional responses. Church services were dramatically taken over as congregants
became infused with Spirit and expressed their joy at this through singing, faint-
ing, running around. On the other side of the Atlantic, the Great Awakening had
its counterpart in the Methodist Church, founded by John Wesley in Bristol,
England, in 1739. Wesley was a profoundly charismatic preacher and brought
profound innovation to British Christianity, taking the church to the new rural
proletariat, preaching in the open air, reducing hardened miners to tears.

There is surely something egalitarian, levelling, in these emotionally charged,
anti-institutional, eighteenth-century movements of the Spirit. The message that
'God is no respecter of persons' certainly found a voice in its benches and fields.
The movement spoke to the experiences and needs of the most downtrodden
and dispossessed – becoming the churches of miners in Britain, and of slaves
in America (MacCulloch, Bancroft, and Salt 2010). Awakening Christianity was
associated with stoking slave rebellions in Boston and New York, and some in

the American colonial establishment 'feared that the Levellers, Ranters, and Fifth Monarchy men of the seventeenth-century English Revolution had reappeared' in the form of these new evangelicals (Linebaugh and Rediker, 2000: 190–191). In 1774 Wesley published his *Thoughts on Slavery*, which concluded that 'liberty is the right of every human creature as soon as he breathes the vital air. And no human law can deprive him of that sight which he derived from a law of nature (cited in Linebaugh and Rediker, 2000: 296). It makes sense, then, to think of the Awakenings and the Methodists as in some sense a return, renewal or resting place of the Spirit or 'everlasting gospel' of Winstanley, the Diggers, Levellers and Pirates – the spirituality of the monster.

But the relationship between the monster (the population) and the bourgeoisie was itself undergoing reformation at this time. And the egalitarian, let alone commoning, impetus of the spirituality of the monster was not to be relied upon. As Linebaugh and Rediker show, prominent Baptists of Bristol (presumably a part of John Wesley's milieu) had become firmly entrenched in the colonial slaving economy by the end of the seventeenth century. The egalitarian impulse of their religiosity was made to reconcile with the source of their prosperity through the invention and adoption of modern racism (Linebaugh and Rediker 2000: 97–99). For all the egalitarian and social impulses of the Great Awakening and Methodism, it is clear that the newly minted scientific and state racism also found a place on their pews, as did the promotion of commerce, the money economy and colonial expansion. In the 1780s Methodists played a key role in the prevention of slave revolts in the West Indies, and in the 1790s in Virginia many Methodists had backed away from anti-slavery and sought a 'gospel made safe for the plantation' (Linebaugh and Rediker, 2000: 240).

The Methodists were the church of the miners, but they were also the church of mine-owners (Thompson 1963). Methodists were foremost amongst the missionaries of the British Empire at the end of the eighteenth century. In southern Africa they acted as advocates for local black populations against the ravages of colonial commercial exploitation, they provided education and made their mission stations spaces of security for people dislocated by war. But they were also proponents of 'civilisation' with all its cultural-racist implications, and of commerce, celebrating the money economy and not only the 'work ethic' but specifically work for a wage, going so far as to welcome dispossession because it forced people into supposedly salvific wage labour (Comaroff and Comaroff 1991).[4] The seemingly contradictory impulses of the politics of Awakening Christianity are in line with the dualistic, incorporating and fragmenting political technologies that were in emergence at the time and would come to define modern power relations.

If the seventeenth-century transatlantic bourgeoisie saw its task as the suppression of the rebellious heads of the population (come 'Hydra'); by the nineteenth century the task had become the cultivation, maximisation and exploitation of the vitality of the population (come 'Society'). The life of the monstrous population and its common-ist spirituality were becoming incorporated into an emergent biopolitical imaginary – in which the collective body (race, class, nation) and evolutionary conceptions of nature, life, culture and economy became 'the common' and the centre of spirituality.

As mentioned above, Millner (2015) discusses the bourgeois commons movement of nineteenth-century England – the Open Spaces Movement – to which Octavia Hill was central. Hill is a crucial figure in the history of the modern British state. Not only did she work for the preservation of parkland and co-found the National Trust, as discussed by Millner, she also established novel forms of social housing, was a key architect of the emergent arts of governance of public health and is recorded in the histories of social policy as a pioneer of the British welfare state (Fraser 1973). Against a dichotomous reading of history that will associate commoning only with working class heroes, Millner maintains that bourgeois philanthropists including Hill were genuinely inspired by and working for the commons. From this she argues that the history of the idea of the commons is more complex and contested than people's histories often allow, and that realities of commoning are constituted through aesthetic frames that are always in tension. However, the convergence of bourgeois welfare-state building and the spirit of the commons to which Millner draws our attention might point to more than the complexity and plurality of political history.

Following Michel Foucault, we have come to think of the eighteenth- and nineteenth-century Western European bourgeoisie as having 'invented for itself a class body' (Foucault 1978), defining itself as a class in terms of physical health, vigour and creative capacity to transform the future – a vitality that is understood as being constituted collectively as a class, and being subject to the threat of degeneration through abnormal behaviour and contact with infection. Bourgeois projects of health campaigns, regulation of sexuality, urban reconstructions, public schools – all can be seen as originating in the efforts of the bourgeoisie to manage and care for its own corporate body. This collective body became a common object of passionate spiritual and political devotion and a frame of reality, a 'quasi-transcendental' (Foucault 2002, 2007, Deleuze 1988). The 'discovery or invention' of the collective embodiment of people – the modern biological concept of 'population' – was such a big event in the organisation of perception and care (according to Foucault at least) that it formed the basis of a whole new rationality and organisation of power – proliferating political economy, linguistics and evolutionary biology, future-orientated values and intensive practices of collectively caring for, maximising and exploiting life (Foucault 2007, Blencowe 2012). Foucault contends that the outsiders of this class body demanded and fought for inclusion through countless specific struggles – Chartism and the Labour movement being obvious examples – and that it was through such struggle that the net of collective body production and regularisation was extended to incorporate the proletariat (Foucault 2002). The imaginary of the population then extended from the boundaries of class to become the modern idea of the nation (Blencowe 2012). The collective embodiment of class, nation and race is the site of commonality – the domain of common thing – that sustain the spirituality of modern political ideologies including liberalism, socialism, nationalism, totalitarianism and imperialism.

Reading this moment back through the lens of the spirituality of the commons and Linebaugh and Rediker's hidden history of the revolutionary Atlantic, we might posit a different agency in its creation. When the bourgeoisie 'invented for itself a class body', we might suppose that the new bourgeois sciences (hewn in the corpses of colonies and slums) were not the only source of its learning. Perhaps the bourgeoisie was also learning from the commoning practices and spirituality of the 'monster' that it had spent the past two centuries fighting – appropriating the impersonal common life force that Winstanley and Wedderburn named 'the Holy Spirit', and reproducing this as the impersonal common life force of the bourgeois class body, race, society and nation.

The monstrous corporate body 'with so many heads that it cannot be counted' – the Hydra – reappears in the nineteenth-century bourgeois imagination as the biological population which it will aim to profitably manage (Hinterberger 2012). The common land become-Holy-Spirit of proto-Pentecostal Pirates has undergone a new transubstantiation to incarnate as the intergenerational, intimately and dynamically related, collective, incorporated vitality of the biological population – articulated in the ideas of class, race, society and nation. If this is right, then nineteenth-century bourgeois commons movements were not 'just another version' of the politics of the commons popping up at a different moment in history, but were something more like the incorporation – in-corporeal-isation – of the commoning spirit of the trans-atlantic proletariat. The egalitarian impulse of that spirit re-emerged in the political demands of socialism. But the idea and experience of the collective biological bodies also fostered the most rank exclusions, enclosures and exploitation.

Conclusion: the 'spirit of 45'

It's the evening of 27th July 2012 and I sit amongst friends watching the television screen. I don't know how this has happened – I swore to myself (and anyone who cared to listen) that I wouldn't be watching this: the London Olympic Games Opening Ceremony.

I am the rebellious daughter of a sports fanatic mother – I can't stand spectator sports. I am a feminist, socialist, internationalist and when I see symbols of nations – especially this British nation – I see naught but oppression and violence. And when I see a spectacular media show carved in the carcass of East London I feel fury at all those brilliant unfunded projects and neighbourhoods trashed. And I never watch television anyway. How can I be here watching this?!

Ah the friends cooked me dinner, I'm at their house, they want to watch it, and my lift is yet to arrive . . . the car driver is blatantly watching.

So I watch.

And I am knotted inside myself. Cringing in advance at the clichés and jingoistic triumphalism that I know to expect. Rock-hard fast in my cynicism and indifference to the ploys of this turbo engine of affection.

Commentators let us know that most of the dancers are not professionals but are in fact everyday public sector workers who have been invited to volunteer. A green and pleasant pastoral immediately gives way to an industrial land-scape showing workers fuelling revolutionary technical innovation . . . And I am fully fledged in my icy outlook - sneering at the romanticisation of hard-ship, the celebration of that artless Engineer Brunel, and the exclusion from the vision of the Empire that made Britain (supposedly so) Great . . .

The Queen jumps out of a helicopter with James Bond. All are revelling in the carnivalesque and surreal . . . Still I do not give an inch.

And then.

And then. Nurses flood the stadium. Real Nurses beautiful and dancing in a stadium watched by hundreds of millions. An almighty moment in the sun for denigrated workers . . . And eek I feel myself softening . . . just a little . . . as they make bedframes perform a comic turn . . .

And just when I'm thinking that I can pretty much hold out, that they defi-nitely still haven't 'got' me . . . A giant shining NHS symbol appears in the middle of the stage. Like the whole of the stadium, and the whole of the watching world, is worshipping the National Health Service.

They got me.

Even me.

I melt.

Ken Loach's 2013 documentary *The Spirit of 45* is an attempt to capture the intense spirituality, love and passion that founded the NHS and other institu-tions of the welfare state in post-Second World War Britain. The spirit, or spirituality, is portrayed through celebratory archive footage and through inti-mate personal narratives of 'ordinary working people' as well as the new pub-lic sector workers. A doctor describes with pure joy the day when he was first able to say to the mother of a sick child who could not afford his treatment: 'don't worry about that – from now on the treatment is free'. A former miner describes the squalid living conditions of his childhood in a slum Victorian terrace and our lungs heave in empathetic frailty at the obvious physical costs that life would bear – we sigh out our relief as we learn of the really decent council housing that was to follow and recall the sure hand of that doctor who could now heal the wounds for free. We see love and wonder in the eyes of our elders and understand that the creation of the NHS was truly a spiritual, spirit-full, project. Indeed, we reflect, love for this institution must be the closest thing the British have to a national religion, even if we lack the fervour today to fight for our gods.

And of course the film itself is an attempt to somehow recall and revitalise that spirit. To pass on the word from a generation that is itself passing – to declare that miracles do happen, that new worlds can be made to rise, and that the people of Britain have built and to some extent still possess the most miraculous commons, most purely symbolised in 'our NHS'.

An aspect of the story that the film shows, but seemingly despite itself – that is downplayed in the narration – is the centrality of the experience of war to the generation of this nationalising spirit. The 'spirit of 45' emerged from a concrete experience of collective, national, embodiment. The people of Britain had lived through intense mortal threat that was specifically collective. Limitations and the state control of usually privately abundant resources had made the materiality and interdependence of lives in a territory far more readily perceptible. Enemy lines and coping strategies had proliferated ideologies of 'them 'n us' – from the violent, triumphalist racism engendered in Winston Churchill to the internationalist, socialist patriotism of George Orwell.

The passionate intensity and spirituality that created and maintained the miraculous commoning of the NHS sprang from the soil: the garden soil in which women had been 'digging for victory', and the blood-drenched soil of battlefields in which intense corporeal interdependence and affection bound bodies, technologies and future lives. The commoning spirit is nourished by intense experiences of material ecology.

> It's the next evening – 28th July 2012 – and I am sharing a different meal with different friends: British intellectuals and devotees of the Left. They are elated by the Opening Ceremony – by the inclusion of all these leftist themes in this national spectacle. Voldemort terrorising the children of Great Ormond Street Hospital was, they explained, health secretary Andrew Lansley terrorising the NHS. Mary Poppins defeating Voldemort was a hopeful premonition of leftist revival to come.

> 'And wasn't it beautiful to see the celebration of working people, rather than "National heroes" at the centre stage?'

> Whilst I can't exactly disagree my feelings are very different to their elation.

> I feel dirty and manipulated – 'they even got me!' – exploiting the NHS and workers histories to draw the affections of the feminist, socialist, internationalist into the celebration of machismo, Nation, and spectacular consumption. I feel sullied. Perhaps I feel that something sacred has been ignominiously profaned. I definitely feel like a shmuck, like I've been hoodwinked into feelings against my better nature.

> But much worse than this is the realisation that in this ideological manipulation I am in fact facing the truth: That of course all along the NHS, our passionate common, has been a snare; a hook, a trade-off, securing the attachment of folk like me to the racist, imperialist, capitalist state that I know to abhor; to the profit motives of Big Pharma and military hardware; to the vicissitudes of normalising life.

But in the sickening feeling is also wretchedness in the knowledge that whatever the cynical implications or motivations, I couldn't choose for us to be without the NHS, and that this common does not look set to survive the current wave of enclosure.

The sound of feathers surges in my throat.

I love this *Spirit of 45*. How can I *but* love this spirit that has carried me and mine through sickness, poverty and ignorance – the spirit of the monster out of place. But its egalitarian impulse and generosity are not to be relied upon. Incarnated in blood-bonds and security threat, it all too easily takes horror stories and the splitting of skin for acts of redemption. Ripped from the materiality of the common world, this spirit seeks expression in the intensity of other matter. Invested in the materiality of collective bodies, it renders lived bodies intensive sites of concern – concern that becomes the desire to control, extract and eliminate as readily as it does that to foster. Passionate intensity without the spacing of solid matter to hold the tension in place can collapse into hunger for pure life, the new, exhaustion. The drive for pure life is always also the drive for exposure to death. War drums beat the heart of the collective body. My hope for common-ist politics today is that the reinvestment of spirit in worldly matters of place – in the ecological pragmatics of collaborative property distribution, creation and interconnection rather than the processuality of infinite growth, wealth and health – will conjure rhythms, stories and songs that drown such Sirens.

Notes

1 Reprinted from *Borderlands/La Frontera* (Anzaldúa 1987).
2 References to William Yeats' poem 'The Second Coming' are commonplace (for example Žižek 2015). The poem was written in the aftermath of the First World War and is taken as an ominous premonition of the movements of fascist passionate intensity that were soon to follow. It has become something of an anthem for critiques of modernity, capturing a sense of the tragedy (or spiritual crisis) in which – it is said – that liberal indifference and indecision abound alongside totalising and violent fascistic tendencies. The famous first stanza reads:

> Turning and turning in the widening gyre,
> The falcon cannot hear the falconer.
> Things fall apart; the centre cannot hold;
> Mere anarchy is loosed upon the world
> The blood-dimmed tide is loosed, and everywhere
> The ceremony of innocence is drowned;
> The best lack all conviction, whilst the worst
> Are full of passionate intensity.

3 As we know, commons are frequently incorporated into money making-ventures and capitalised and important arguments exist concerning the dialectical co-constitution of capitalism and the common (Hardt and Negri 2000, 2009). However, I do not wish to replicate the dialectical analytic of capital versus the commons, or the idea of incorporation, which tend to suppose a false opposition between power, on the one hand, and

commoning, on the other, as well as a capitalocentric analysis (Gibson-Graham 2006). As Bresnihan and Byrne (2015) explain, commoning practices generate and confront their own problematics of governance and power. Power, capture, heaving into being, are crucial for all forms of community building and empowerment (Pearce 2013). 'Capture', then, is not the evil other of a 'liberating' force of the commons. In using this term my intention is to point to the significance of spirituality in the creation of spaces and social forms ('social reproduction' as Bresnihan and Byrne might put it) as well as to the perpetual ambiguity, angst and frailty with which such forms and their norms are imbued.

4 John Wesley himself was a great advocate of the salvific potential of money economy, stating: 'Money is of unspeakable service to all civilized nations in all the common affairs of life. It is a most compendious instrument of transacting all manner of business, and (if we use it according to Christian wisdom) of doing all manner of good. [It is] . . . food for the hungry . . . raiment for the naked . . . [and] father to the fatherless' (cited in Comaroff and Comaroff 1991: 170).

References

Anzaldua, Gloria. 1987. 'Borderlands/la frontera.' San Francisco, CA: Aunt Lute.

Arendt, Hannah. 1958. *The human condition.* 2nd edn. Chicago: University of Chicago Press, 1998.

Beckford, Robert. 1998. *Jesus is dread: Black theology and black culture in Britain.* London: Darton, Longman & Todd.

Blencowe, Claire. 2012. *Biopolitical experience: Foucault, power and positive critique.* Basingstoke: Palgrave Macmillan.

Blencowe, Claire, Brigstocke, Julian, and Tehseen Noorani (2015) 'Engines of alternative objectivity: re-articulating the nature and value of participatory mental health organisations with the hearing voices movement and stepping out theatre company'. *Health*: An Interdisciplinary Journal for the Social Study of Health, Illness and Medicine.

Bresnihan, Patrick, and Michael Byrne. 2015. 'Escape into the city: everyday practices of commoning and the production of urban space in Dublin', *Antipode* 47(1): 36–54.

Comaroff, Jean, and John L. Comaroff. 1991. *Of revelation and revolution, volume 1: Christianity, colonialism, and consciousness in South Africa.* Chicago: University of Chicago Press.

Deleuze, Gilles. 1988. *Foucault.* Minneapolis: University of Minnesota Press.

Deleuze, Gilles, and Félix Guattari. 1988. *A thousand plateaus: Capitalism and schizophrenia.* London: Bloomsbury Publishing.

Foucault, Michel. 1978. *The history of sexuality: An introduction. Vol. 1.* New York: Vintage.

Foucault, Michel. 2002. *The order of things: An archaeology of the human sciences.* Oxford: Routledge.

Foucault, Michel. 2007. *Security, territory, population: Lectures at the College de France, 1977–78.* Basingstoke: Palgrave Macmillan.

Fraser, Derek. 1973. *The evolution of the British welfare state: A history of social policy since the Industrial Revolution.* London: Macmillan.

Freire, Paulo. 1970. *Pedagogy of the Oppressed*, trans. Myra Bergman Ramos. New York: Continuum.

Gibson-Graham, Julie Katherine. 2006. *The End of Capitalism (as We Knew It): A feminist critique of political economy; with a new introduction.* Minneapolis: University of Minnesota Press.

Hardt, Michael, and Antonio Negri. 2000. *Empire*. Cambridge, MA: Harvard University Press.

Hardt, Michael, and Antonio Negri. 2009. *Commonwealth*. Cambridge, MA: Harvard University Press.

Hinterberger, Amy. 2012. 'Investing in life, investing in difference: Nations, populations and genomes', *Theory, Culture & Society* 29(3): 72–93.

Honig, Bonnie. 2012. 'The Politics of Public Things: Neoliberalism and the Routine of Privatization.' *No Foundations: An Interdisciplinary Journal of Law and Justice* 10: 59–76.

Katongole, Emmanuel. 2011. *The sacrifice of Africa: A political theology for Africa*. Cambridge, UK: Wm. B. Eerdmans Publishing.

Linebaugh, Peter. 2014. *Stop, Thief! The Commons, Enclosures, and Resistance*. Oakland, CA: Pm Press.

Linebaugh, Peter, and Marcus Rediker. 2000. *The many-headed Hydra: Sailors, slaves, commoners, and the hidden history of the revolutionary Atlantic*. Boston, MA: publisher.

MacCulloch, Diarmaid, Gillian Bancroft, and Siân Salt. 2010. *A history of Christianity*. London: BBC Worldwide.

Pearce, Jenny. 2013. 'Power and the twenty-first century activist: from the neighbourhood to the square', *Development and Change* 44(3): 639–663.

Read, Jason. 2011. 'The production of subjectivity: from transindividuality to the commons', *New Formations* 70(1): 113–131.

Taylor, Mark Lewis. 2004. 'Spirit', in *Blackwell Companion to Political Theology*, Peter Scott and William Cavanaugh (eds). Oxford: Blackwell.

Thompson, Edward Palmer. 1963. *The making of the English working class*. London: Victor Gollancz.

Werner, Dietrich. 2013. 'Giving glory to the God of life in the context of 21st century christianity.' In Edison Kalengyo (ed.) *God of Life*. Kenya: All Africa Conference of Churches.

Žižek, Slavoj. 2015. 'On the Charlie Hebdo massacre: are the worst really full of passionate intensity?' *New Statesman*, 10 January.

10 Controlled natures

Disorder and dissensus in the urban park

Samuel Kirwan

Introduction

Not only for landscape designers (Cranz, 1982), historians (Conway, 1991, 2000) and observers of public health, both mental (Barton and Pretty, 2010; Bowler *et al.*, 2010) and physical (Bird, 2004; Wheater *et al.*, 2007), a walk in the park has in recent decades been enjoyed by cultural and moral geographers (Lawrence, 1993; Matless, 1997), scholars of race and ethnicity (Byrne and Wolch, 2009) and non-representational accounts of visibility (Wylie, 2002) and alienation (Olwig, 2005). As the urban park has also, looking to Gezi and Zucotti parks, become the focal point for forms of protest that have disrupted established rhythms of collective political action, their contours are being newly mapped by social movement theorists (Langman, 2013; Tufekci, 2013) and political scientists (Arat, 2013; Kuymulu, 2013) interested in the entanglements of politics, space and experience that constitute these specific historical moments.

For this walk can be an instructive trip. At the heart of the city, the park presents a negotiated abundance of nature, a reimagining of the rural landscape mediated by the complex ecology of the city (Gabriel, 2011). The urban park is *release* – from life saturated by economic, temporal and spatial pressures – yet there is also a wealth of aesthetic sadness in the urban park, a *failure* to live up to a certain imagination of 'nature' as perfectly and pleasingly *uncontrolled*. The park harbours extremes of both order and disorder; on the one hand, it is the shocking sight of a group of youths lounging amidst piles of rubbish, shrouded in loud music and aggression, or the fear that takes hold at night in the absence of street lights and CCTV networks. On the other, the verdant lawns and manicured flower-beds are lamented by historians and social scientists for their excesses of control (Marne, 2001; Taylor, 1995; Firth, 2003). The municipal park emerged in a period of the rapid enclosure of 'the commons' as a space that is bordered, managed, policed, guarded, owned and observed; it has long been associated with the loss of the truly 'open' commons. Since the mid-1990s these same spaces, many rejuvenated through Heritage Lottery Fund grants, have been subject to new forms of surveillance and banishment.

Approaching these tensions between containment and opening through the work of Jacques Rancière, I seek in this chapter to give a novel reading of the role of the urban park in its relationship with this romanticised 'commons'. The chapter argues that if 'the commons' is less a state to be regained than an opening to equality inscribed in space, then, in contrast to the narrative placing the park as indicative of the 'enclosure' of the commons, the park may be seen as indicative instead of methods for introducing the commons into the city.

This argument rests on the unusual approach Rancière takes to the concept of the aesthetic, one that forces a questioning of how the social sciences approach aesthetic experience. Having examined the 'sociologisation of the aesthetic' that frames critical perspectives on contemporary experiences of public space, the chapter proceeds to set out the two principal designations, as I see it, of green space as harbouring politics of enclosure. The first is the enclosure of experience: the shaping of the wild nature of the park with the appetitive pleasures of the working-class subject. Examined are historical accounts of the emergence of the first municipal parks and the park movement, paying particular attention to the Select Committee on Public Walks, whose presentation to Parliament marked a pivotal moment in the birth of the free-to-access park (Conway, 1991). The second is the enclosure of shock: the distribution of disruptive bodies, behaviours and objects that seeks to avoid any disturbances in the aesthetic expectations of a given space.

However, the chapter follows Rancière in identifying the key *political* moment to be not these logics of enclosure but, rather, the suspensions and disturbances to these forms of containment. Rancière enjoins us to consider the aesthetic not only as experience, but also as *suspension*: the suspension of our particular position that, as determinate subjects, we experience in an aesthetic relation. In a final section, having introduced the basics of Rancière's theoretical framework, the chapter notes the multiplying inconsistencies, suspensions and reappropriations of these forms of enclosure and develops a critique of the notion that these logics of containment are *all there is*.

Aesthetic commons

A principal reason for considering aesthetics in the context of urban space is the rise of a more strategic, wide-ranging and invasive policing of what we are likely to see (Deleuze, 1992; Rose, 1999). Like other urban spaces serving as platforms for diverse uses and experiences, since the dawn of the 'New Labour' era public parks have been subject to an increasing range of techniques, brought together under the term 'anti-social behaviour', designed to limit, disperse or discourage instances of aesthetic shock (Burney, 2005; Prior, 2007; Squires, 2008). In sum, a range of analyses have made abundantly clear the extent to which the aesthetic is the site of politics and power. Made clear in these developments is the extent to which, when discussing aesthetics, we are discussing *community*. To consider the beauty of a flower-bed or verdant lawn is to recognise the level of *correspondence* between our own aesthetic experiences and those of others, our belonging to a community of users both past and present. Correspondingly, it raises the limits of this togetherness: why is this experience not respected by others – the motorbike rider and fly tipper, the tagger or arsonist?

In discussing the aesthetic and commonality as linked in this way, we are raising the *suspension* of our particular selves. Rather than individuals engaged in an experience determinable by our own composition in time, space and society, we are instead considering a *suspension* of who we are as bodies composed by a connection with objects that cuts across individuals. Again, there is a good reason for considering this suspension to have political dimensions. The Occupy movement in New York, London and elsewhere presented the gesture of suspending the predetermined uses and audiences of public space as a *sui generis* political act. To engage in action with no predetermined political goals or outcomes, only the aesthetic act of creating a new distribution of roles and experiences, becomes meaningful in an economic and political climate that appears beyond any meaningful change (Žižek, 2012).

These developments chimed with the concept of 'the commons' (or 'the common'), a concept that has been seized upon by several contemporary thinkers (see Hardt and Negri, 2009; Linebaugh, 2009; Gibson-Graham and Roelvink, 2010) as an evocation of commonality and sharing without the implications of consensus and communion denoted by the term 'community'. The tents and inchoate forms of action stirred thoughts of pre-enclosure spaces and the promise of a more equitable and egalitarian organisation of economy. What I seek to do in this chapter is to display how, in considering the commons as this aesthetic suspension, it may be presented less as a specific historical space to be recreated than as a disruptive moment that suspends an established distribution of acceptable behaviours. I argue, finally, that the public park, with nature as play and experimentation, has been a privileged site within the urban environment for this disruptive moment.

The sociologisation of the aesthetic

The social sciences have tended to reject a romantic approach to aesthetics (Armstrong, 2000) in which aesthetic experience, as wonder and awe, as both originary connection to nature and transcendent connection to the divine, would to some extent lie outside of an individual's social position or determinable capabilities. To foreground the primacy of the social, on the other hand, is to approach such questions in terms of different aesthetic capabilities. The seminal text, when considering this rejection, is Pierre Bourdieu's *Distinction* (1984). Bourdieu demonstrated how individuals' reactions to artistic objects were determined by their socio-economic background. When given an abstract photograph, working-class participants would react to the *reality* of a photograph, judging whether or not it fulfilled a certain determinable function. Such individuals 'lacked the specific competence' (44) to understand 'artistic' photos, whether 'abstract' or of unusual subjects, and were eventually forced in these cases to 'admit defeat' (46). That is, forcing them to recognise their own non-understanding, inasmuch as bourgeois photography was accepted to be beyond their frames of comprehension, such moments demonstrated to these subjects the reason for their social exclusion. Thus described by Bourdieu is the aesthetic as an operative fiction, distributing places within society, maintaining the illusion of the disinterested, super-sensible experience serves to display to individuals the necessity of their being dominated within this structure. In short, because they do not understand bourgeois photography, they must remain in the subordinated roles of manual labourers. Following De la Fuente (2000), we may label this approach, through which the sociological understanding of the world proceeds by delimiting aesthetic judgement in line with societal categories, the 'sociologisation of the aesthetic'.

Described in the sections below are critical approaches to the creation of public parks – critical, that is, of the liberal narratives that celebrated their democratic, equalising and educative qualities – that adhere to this formula of demarcating aesthetic capabilities in line with societal positions. As will be discussed, these identifications of an enduring logic of *enclosure* that characterised the Victorian public park rely upon a clear demarcation of the aesthetic capacities of the working classes. These perspectives make clear the implications for our contemporary use of green space; the implications of freedom and pleasure hide the existence of rigid symbolic codes regarding acceptable behaviour and subtle barriers that marginalise and exclude certain social groups. In presenting, through the work of Jacques Rancière, a challenge to this hierarchical superiority of the social with regard the aesthetic, the chapter also presents a challenge to the reading of urban parks as enduring spaces of 'enclosure', and seeks to do this through that most evocative area of behaviours: disorderly or anti-social conduct. Thus, before moving on to examine these critiques and a 'Rancièrian' response to them, it is necessary first to establish how this 'sociologisation of the aesthetic', and a focus upon the enclosure of experience, continues to structure social scientific approaches to 'disorder'.

Anti-social aesthetics and the enclosure of shock

As New Labour came into power in 1997, amidst millennial talk of 'Cool Britannia' and a new multicultural settlement (Modood, 2007: 10), the rather more murky business of 'anti-social behaviour' was widely agreed to be the most pressing political issue facing the nation (Burney, 2005: 1). Critics (Ashworth *et al.*, 1998; Squires, 2007) of anti-social behaviour policy implemented by New Labour have focused particularly on the claim that, rather than serving any rehabilitative purpose, Anti-Social Behaviour Orders (ASBOs) and the other interventions introduced to combat the plague of anti-social individuals indicative of 'Broken Britain' were being used by the police to *manage*, rather than rehabilitate, difficult populations. Furthermore, rather than strengthening community through the fostering of collective action, in tending towards enforcement-based practices and a notion of community based upon a morally virtuous subjectivity, they fostered instead the exclusion and marginalisation of significant sections of the working poor (Prior *et al.*, 2006).

Andrew Millie's (2008a, 2008b) work is exemplary for our purposes here, since it clearly articulates the relationship between the aesthetic and anti-social behaviour that prevails within these critical approaches. This approach can broadly be divided into a 'Bourdieusian' stage and a 'Foucauldian' stage: first a mapping of particular tastes and judgements particular to individuals and spaces, and second a mirroring of this mapping in spatio-temporal techniques for the distribution of aesthetic pleasures.

Millie proposes that our socially specific tastes, and the particular spatio-temporal context in which these are deployed, underpin any attribution we might make of a behaviour's being 'anti-social' (2008a: 384). As a critical approach it presents a noble attempt to return the material to policy analysis: in other words, to recognise the situation of events in times and spaces and to recognise that nothing occurs abstractly, but in a particular social situation. In line with Bourdieu's work, Millie's project is to demystify, through sociological categories, the 'objective' claims that underpin certain aesthetic judgements, and is as such based upon the bracketing out of the aesthetic as this suspension of the purely subjective.

The example Millie gives of the concrete situation of judgements, one frequently used in contemporary discussions of the relativity of anti-social behaviour, is graffiti, in particular with reference to its turn, as a cultural artefact, from mindless defacement to, in selected circumstances, bourgeois respectability. Thus, Millie displays how the distribution of 'acceptable' and 'unacceptable' graffiti, between 'spray art' and 'tagging',[1] relies upon shifting interpretations of what is acceptable within urban space. Such a relativity is adequately displayed, as Millie notes, in the elevated position, among middle-class observers and Bristol City councillors, of the graffiti artist Banksy (Millie, 2008a: 386).[2]

Following Bourdieu, Millie assumes that the aesthetic claim to universality displayed in these judgements, that in which one graffito is celebrated as art while another is denounced as vandalism, is illusory. In the latter case, it follows that if a tagger takes pride in their work and sees in it a certain beauty, while local residents see only the beauty in the park bench it is scrawled on, then all are only expressing their particular composition as subjects in determinate situations. The line dividing these judgements of taste is purely social; their mutual claims to a certain objectivity are illusions within a constructed matrix of perception. Our perception of an object as beautiful or anti-social is only the expression of our situated, enculturated judgement.

As noted above, the second stage of Millie's analysis, in which he draws our attention to how this matrix of judgements, individuals and spaces operates as a locus of power, owes less to Bourdieu than to Foucault and Deleuze. This analysis builds particularly upon a group of authors (Rose, 1999; Dean, 1999) working with Foucault's short lecture on 'governmentality', whose work has brought to light the spatial techniques for the *distribution* of gazes and activities that create certain aesthetic ecologies. The wish to shop, work, live or play in a pleasant environment is itself the site through which various techniques, such as the ASBOs and Dispersal Orders, are both enacted and legitimated. It is in this vein that critics of anti-social behaviour policy have noted the extent to which preventative interventions largely abandon 'rehabilitative' procedures in favour of behaviour management (Chakrabati and Russell, 2008). A particular manifestation of this is the tendency of the police and other 'enforcement' agencies to use ASBOs and other interventions in order to 'cleanse' certain spaces of the 'usual suspects', namely sex-workers (Hubbard, 2004; Sagar, 2007; Scoular *et al.*, 2007), drug users and other 'street-life' individuals (Moore, 2008) and young people (Goldsmith, 2008).

In other words, aesthetics is inscribed in space not only in socially and historically conditioned individual determinations of 'beauty', but through the varying aesthetic techniques that compose the 'urban' in its material complexity. As Millie notes:

> In effect, the untidy are removed or hidden from view so as to beatify the city. 'Popular' aesthetics are catered for in the creation of a safe and sanitized streetscape, acceptable to the shopping, business, leisure and residential majority.
>
> (2008a: 387)

Described here is the material operativity of the aesthetic. For, on this account, not only are there clear societal demarcations regarding who may experience what, but there are also structures of power working to maintain these distinctions. We might observe how certain urban spaces, of which shopping centres and high streets are the most prominent, are increasingly permeated by techniques, operating upon bodies, spaces and times, that serve to eradicate any disruptions to the pleasant aesthetic in which, as Millie recognises, many of us wish to live. The ASBOs and No-Drinking Zones, we realise, are all part of a system seeking to keep individuals' aesthetic experiences in their proper places. It is here that the parallels with 'enclosure' as a historical process become clear; the sociologisation of the aesthetic, in which individuals' aesthetic capabilities are enclosed within clear categories, is mirrored by a *regime of enclosure* that seeks to maintain these distinctions.

As I read Rancière's work, he does not seek to contradict or falsify these latter forms of power. He *does*, however, challenge the notion that the mapping of aesthetic capacities is *all there is*. He claims that to subscribe to these orderings of experience is to continue their policing of experience; the regimes of enclosure, maintained through techniques of behaviour management, *rely upon* an unquestioning acceptance of these demarcations of proper aesthetic capacities. As we turn now to how critical accounts have portrayed the enclosure of experience in supposedly wild and freeing natures of the Victorian urban parks, it is worth repeating the argument of this paper: that the truly critical gesture is not to reveal these structures but, rather, to bring to light the multiple moments in which they are transgressed; it is to display the contingency of the story that is told about those who are unable or unwilling to partake in the 'high' aesthetic pleasures reserved for the privileged.

The loss of the commons and the enclosure of experience

> They hang the man, and flog the woman,
> That steals the goose from off the common;
> But let the greater villain loose,
> That steals the common from the goose.
>
> (Anonymous poem of seventeenth century)

Though cities had private gardens and commons, and London had the Royal Parks, which had been made progressively open to the public over the seventeenth and eighteenth centuries (Conway, 1991: 12), the concept of the publicly accessible, municipally funded park is relatively recent, dating to the significant urban expansions of the mid-Victorian era. Yet, on many accounts (Marne, 2001; Firth, 2003; Taylor, 1995), the urban parks that emerged in this latter period simply *continued* the theft, described in the well-known anonymous poem of the seventeenth century, whose effects they were intended to mitigate. Highlighted in these accounts is the extent to which these emerged less from a concern for public enjoyment than from paternalist fears regarding the moral health of the 'humbler classes' and a more visceral fear of the popular movements, particularly Chartism and the Reform League, for which public spaces played both a symbolic and a practical function (Roberts, 2001). These critiques present a specific case study within the wider argument that the middle and upper classes in the Victorian era maintained their wealth and power against a backdrop of material inequality, not through force but by the imposition of middle-class values (Stedman-Jones, 1971; Donajgrodzki, 1977), in this case the genteel qualities associated with a gentle stroll.

Indeed, when Parliament was presented in 1833 with the Report from the Select Committee on Public Walks (hereafter SCPW) (Slaney, 1833), extensive focus was placed upon the new working-class subject that would emerge from the greater provision of green space:

> A man walking out with his family, among his neighbours of different ranks, will naturally be desirous to be properly clothed, and that his Wife and Children should be also; but this desire duly directed and controlled, is found by experience to be of the most powerful effect in promoting Civilization, and exciting Industry.
>
> (Slaney, 1833: 9)

The 'park movement', of which the SCPW may be seen as a founding moment (Conway, 1991: 21), was not only born of the pace of enclosure and 'problem' of recreational space in England's urban centres, but also of the embarrassing visibility of working-class recreations; the 'drinking houses, dog fights and boxing matches' (Slaney, 1833: 8); pastimes to which men were 'driven' by the paucity of options for more sedate recreation available to them. The reformers arguing for greater access to parks were part of the wider movement, driven by a belief that leisures and pastimes should be sites for improvement, to foster 'rational recreation' (Bailey, 1987). Another factor to be considered is the depth of an anti-urbanism among the wider reform movement, manifested in a belief that the city had a deleterious effect upon morals (Rosenzweig and Blackmar, 1992; Sennett, 1992), and as such the imperative that 'breathing spaces'– a metaphor with both societal and physical resonance – be integrated into the planning of the city (see Loudon, 1833: 698). As well as being the 'lungs of the metropolis' (Slaney, 1833: 15), the public park was its *conscience*: a 'machine to fix a broken society' (Young, 2004: 3).

Thus, Anthony Taylor notes the manner in which, although parks were presented as spaces of free enjoyment, unwanted activities – principally public meetings and prostitution, but also gambling, courting and other more innocent pastimes – were 'ruthlessly purged' from the 'imagined utopia' of the park. Taylor continues that the claims to inclusivity advanced by the park reformers were rendered absurd by the preoccupation with shaping the behaviours of the working classes:

> Above all, whilst providing access to new green space, it at the same time sought to regulate and police that space. The urban parks of the 1840s and 1850s were accordingly regulated spheres, scrupulously maintained, and patrolled and policed by the hated park-keepers, who became a part of working-class demonology in their own right, co-operated with the police, dressed like them, and used fences, gates and padlocks to exclude. . . . Far from creating a space in which the classes might mingle, the parks rather reinforced existing social divisions in London.
>
> (Taylor, 1995: 386)

Pauline Marne's study of Sefton and Stanley Parks in Liverpool demonstrates the extent to which claims for inclusivity on the part of the park reformers belied strong symbolic barriers inscribed in the park spaces (Marne, 2001: 438). She notes in particular that, where women were encouraged to use certain park spaces, their doing so was heavily circumscribed, and 'was attained at the expense of the freedom of others' (439), principally working-class men. Marne notes the exclusion of the local Irish residents (who made up a quarter of the late-Victorian population of the city and were concentrated in the poorest neighbourhoods) through the banning of 'anyone "attired otherwise than in a decent manner"'. This symbolic exclusion of ethnic groups raises the extent, more comprehensively examined in the North American context (Byrne and Wolch, 2009), to which the idealised vision of nature was predominantly *white*, premised upon an imagination of a rural British identity unsullied by the heterogenous social texture of the urban.

While the park movement inaugurated by the SCPW was slow to build momentum, by the end of the Victorian era significant areas of land had been saved from enclosure and made available for public use. Yet this escape from enclosure, these critical perspectives highlight, harbours another, more insidious form of containment. As opposed to the unmanaged commons, the landscapes of the new urban parks were carefully managed expressions of a utilitarian reformism enacted through a deterministic vision of nature as a kind of moral and intellectual tonic. Kenneth Olwig (2005: 34) describes this process as the *commodification* of landscape, in which the commons, harbouring shared practices and customs that had developed and co-existed there across generations, were identified as waste to be replaced by a carefully managed space 'estranged from its substantive social meaning, the land of a people as *res publica*'. In sum, the park continues the enclosure of the commons because it performs a kind of commodification of the freedoms experienced in the open spaces that preceded the urban expansions of the Victorian era. Rather than opening the body to new experiences, it encloses the body in fixed rhythms. Rather than opening forms of aesthetic pleasure, it inscribes into green space fixed languages of ownership. The creativity, contingency and openness to others that ran through the commons are entirely closed off in favour of strictly managed common body.

In the third section we will address the problems with this approach, focusing in particular upon the assumption that the aesthetic experiences of the working classes who were enclosed in this way can be determined within such an immutable hierarchy. Before turning properly to this Rancièrian critique, it is worth sketching the similarities of approach between the critical observations on the management of behaviour and the emergence of the municipal urban park, inasmuch as both discourses ascribe to a 'sociologisation of the aesthetic'. The implicit assumption I wish to highlight here, for reasons that will become clear below, is that this distribution of experiences maintained by a structure of domination is itself an accurate reflection of the distribution of sensual capacities between groups. As stated above, I wish to question the assumption that this distribution of experiences is *all there is*.

Returning to contemporary anti-social behaviour interventions, Millie's work describes how these serve to demarcate and construct a system of regulations and enforcements based upon perceived differences between aesthetic life-worlds. Yet, in this irresolvable difference there is the assumption that, say, the working-class populations consumed parks only in the terms laid out for them by the reformers in the park movement, or that the group of teenagers who are removed from a green space through a Dispersal Order cannot experience it with the same aesthetic pleasure as do the community group who garden it. It is the assumption that the suspension of the cognitive faculties in an aesthetic experience, the manner in which our determinate modes of cognition are *transcended*, is an ideological illusion and may be fully explained by our background and the time and space in which we are composed. Marginalised, in other words, are the complex languages through which young people in particular articulate their own attachment to particular park spaces.

Precluded in these two accounts of the aesthetics of the park, the first charting the emergence of the park as *enclosure* of experience, the second as an enclosure of shock, is any permeability regarding the boundaries of experience. They preclude any possibility, that is, of the distribution of capacities being transgressed in an aesthetic encounter. What the critical accounts of the emergence of public parks struggle to incorporate is the proposition that working-class populations might have created their own aesthetic languages for enjoying the manicured parks and ordered flower-beds, or that young people inhabit complex ways of owning and making sense of spaces of which, it is often the case yet rarely noted, they are the most frequent users. Asserted in these accounts, that is, is the *political* claim that the places that a structure of power assigns to certain individuals are the places they inhabit, and that one's relationship to power must begin from displaying to individuals the part that they play within a structure. I argue that such accounts contribute to, rather than disrupt, the spirit of enclosure that they seek to critique. In the next section we turn to Jacques Rancière, in whose spirit these critical observations have been formed, and whose project has been formed upon directly opposing these latter assertions, before setting out how this claim may be challenged.

Rancière and aesthetic equality

Jacques Rancière is part of the generation of philosophers, of whom Jean-Luc Nancy and Alain Badiou would be the other leading voices, whose philosophical trajectories have been shaped by a desire to recapture philosophical questions that are seen to have been lost in the work of Foucault, Deleuze and Derrida. If, for Nancy, this question is community, for Rancière it is democracy. As one of his key introducers to English-speaking audiences explains, '[h]is unflinching defence of a radical version of democratic equality has made him one of the key references in contemporary political thought' (Deranty, 2010: 1). As noted above, the principal terrain of this defence has been the question of aesthetics, and the principal villain therein Pierre Bourdieu. Thus, in setting out this radical articulation of equality, we will finish upon a very different entanglement of space, behaviour and aesthetics than that presented above, one that will open out a different perspective on green space to be pursued in the final section of the chapter. To describe Rancière's unique approach, I will focus here upon a number of concepts that have been explored in his work, among them the 'distribution of the sensible' and the 'uncertain community'. The key terms for this discussion, however, will be Rancière's reframing of 'police' and 'dissensus'. If the techniques of power described above are framed by the former term, the radical value of Rancière's work, as I see it, lies in his taking us *beyond* the critical approaches it names, inasmuch as the moment of 'dissensus', and its centrality to the 'aesthetic regime', form the very specific notion of transformative democratic practice he sets out.

Thus, we will begin with the 'distribution [*partage*] of the sensible', and the multiple meanings the phrase carries. As meaning both to share and to divide, *partager* implies the forms of distribution described above, in which a shared sensual landscape is established by dividing between individuals' specific sensual capacities, including the partitioning off of those who *cannot* understand or perceive in certain situations – a group described by Rancière as the 'part which has no part' (1999: 30). Thus, as Rancière describes it, the term denotes 'the system of self-evident facts of sense perception that simultaneously discloses the existence of something in common and the delimitations that define the respective parts and positions within it' (2004: 12).

In sum, the distribution of the sensible concerns what is able to be experienced, or 'what can be said, made or done' (2004: 85), in certain zones, and the careful management of these dividing lines. The management of this distribution, as the holding of those fragile dividing lines that render some utterances perceivable while others are merely noise, Rancière labels the *police*. As Chambers (2010: 63) notes, while Rancière's concept of the police does offer a general description of the role of the institutional 'police force', in reality a polity with a strong police order will have little use for an actual police force, whose intervention is needed only when the former is disrupted. For 'the police', as the political management of this clear distribution of what can be experienced, denotes a far broader system of containment, of which the sociological accounts in their demarcation of aesthetic capacities play an important role. This is because these accounts, in marking out the possibilities of experience, take only aesthetic *consensus*, in other words, the *affirmation* of the distribution of the sensible, as their material of study. Rancière enjoins us to examine instead how the distribution of the sensible as the *set of possibilities* of sensual experience suggests also the possibility of *dissensus* – of the disruption of this distribution and, as such, of the police order. What Rancière finds disclosed in the work of Friedrich Schiller, and in the 'aesthetic regime of art' of which his work is indicative, is the extent to which what is able to be experienced transcends one's subject position. The distribution of the sensible described by Schiller is less a set of stable divisions than a principle of constitutive indeterminacy, one that renders the allocation of particular aesthetic capabilities to particular individuals fundamentally *unstable*. The aesthetic experience for Rancière is less an expression of taste than this act of *dissensus*, the breaking of that connection between what one is and what one is able to see. The democratic action of aesthetics is to be placed not, as in Bourdieu, in revealing the structure that determines aesthetic capacities, and thereby affirming them, but in *transgressions* of this structure.

With this in mind, we may turn, finally, to community. In *The Politics of Aesthetics* (2004), Rancière notes how certain political acts or literary statements, in practising such a suspension, 'reconfigure the map of the sensible by interfering with the functionality of gestures and rhythms', and may as such be seen as the enactment of 'uncertain communities' (40). For the community, in such practices, would not lie in a consensus or togetherness, but in the very calling into question of the 'distribution of roles, territories and languages' (40) that demarcate the 'parts' or 'lots' that compose the community. What Rancière recognises in the aesthetic regime is a radical mode of practising community, one in which what is 'in common' is not a property or consensus, but the presupposition of an *equality* dissimulated by this distribution of parts.

Rather than the 'uncertain community', I propose that it is more appropriate to refer to this moment of radical equality as the 'commons'. In other words, through Rancière, we may see the commons not as a historically past space but as the suspension of that division separating those who know how to sense, and be active, and those who do not. The commons does not harbour equality as similarity but, rather, equality as a political event in which the presupposition that *all* are the same enacts a reconfiguration of the distribution of the sensible.

Rancière notes how this presupposition, as enacted by the nineteenth-century teacher and pedagogist Joseph Jacotot, disrupts the justification for students' remaining silent, for the democratic presupposition of equality serves only to demonstrate 'the sheer contingency of the order' (1999: 30) that precludes their voices. Thus, in the case of Jacotot's radical pedagogy, the student, whose lot it is to listen, learn and be formed in the image of an imparted knowledge, in short, who is subject to a distribution of roles in which their voice does not count, or emerges only as the senseless babble of anti-social grunts, was able take a 'part' where previously they had no part. The 'commons' is not a thing, but this moment in which a radical presupposition of equality disrupts a determining hierarchy. As Todd May describes it:[3]

> It is not that people necessarily demand equality, or even think of themselves consciously as presupposing equality (although often they do). Rather, it is there, in their political practices.
>
> (2010: 72)

In other words, the presupposition of equality lies in the suspensions and transgressions of the distribution of the sensible by the commons. If, in Jacotot's openness to these suspensions, he created conditions for the enactment of *dissensus* that the commons is, Rancière's political demand lies in the continual re-engagement of this moment. It is a demand directed against the dominant forms of critical engagement with the social, which, as noted above, are described by Rancière as contributing to the *policing* of the distribution of the sensible – in short, to the *neutralisation* of equality. The 'sociologist' (by which Rancière primarily intends Bourdieu) remains bound to the police order inasmuch as he considers his work to be one of *unveiling*, of revealing what was hidden, and, as such, establishes a position of mastery over the subject, who may see and experience only that within the bounds of what their *ethos* allows them. Sociology is, from its formation, 'a war machine against allodoxy', a refusal of 'the dehiscence between the arms and the gaze' (Rancière, 2006: 7).

Parks revisited

This refusal, I argue, has been present in the accounts discussed thus far, concentrating as they did upon the ways in which aesthetic experiences are controlled: how they are *enclosed* in a process mirroring the loss of the physical commons. Written out of these accounts, or moreover included as minor exceptions to the general rule of enclosure, are the manifold ways in which these spaces were, and are, appropriated in diverse and unprescribed ways by local populations. In drawing upon these exceptions to the rule of aesthetic containment, I wish to indicate the presence of the commons, of moments where supposedly marginalised communities create from green space their own aesthetic languages and experiences.

A first critical point to note is the extent to which the identification of the park movement and the birth of the municipal park as the establishment of a certain *order* risks overstating the capacities of the reformers. These critical accounts assume, for example, that the manicured urban parks comprised the only open space available to the working classes. Emma Griffin (2005) notes how little contemporary discussion there was of a lack of open space and the extent of disused space available for various recreations (175) owing to the spatially dispersed nature of growth in England's industrial centres (Trinder, 2000). In other words, the notion that the unmanaged spaces of the commons were fully eradicated, with the manicured and managed parks taking their place, overplays the smoothness of the urbanisation process in its enclosure of the open, the common and the hidden.

This over-estimation of the power of the reformers occludes also the agency of working-class individuals in shaping the new municipal parks. A seminal text in this respect is Roy Rosenzweig's (1985) study of the role of immigrant Polish populations in the shaping of parks in Worcester, Massachusetts. Rozenzweig argues that

> proponents of the social-control formula suggest that the object of reform designs – the urban worker – was both inert and totally pliable. By viewing park reform exclusively from the 'top down,' they ignore the possibility that workers might have taken an active part in conceiving or advocating parks and assume that workers uncritically accepted the park programs handed down by an omnipotent ruling class.
>
> (1985: 127)

In the United Kingdom the role of 'workers' in the formation and shaping of green spaces is poorly understood, principally because the records available – minutes of committee meetings, documents detailing lease agreements and acquisitions, newspaper reports – focus upon the bureaucratic process at the expense of local experiences and forms of expression.

Taking my own city, Bristol, as a case in point, an account of the birth of its major green spaces drawn from such 'official' sources would proceed by detailing the series of leases, gifts and purchases over the second half of the nineteenth century. Bristol lagged behind the other urban centres in the provision of green space, the city corporation being 'slow to anticipate the need for public parks and slow to respond to the public demand for them' (Young, 1998: 182). What such accounts omit, as Young notes, is the consistent pressure that was placed upon the corporation from 'ratepayers, residents or workers for the establishment of a park in a particular district, for the hastening of the acquisition process, or for the provision of some facility within a developing park' (1998: 180). The creation and shaping of several parks attained in this period, Young argues, depended upon the direct but largely unrecorded involvement of local individuals and groups. The clearest articulation of the involvement of the local working population in the case of Bristol is given in the pamphlet, published in 1871, entitled *The Cry of the Poor; Being a Letter from Sixteen Working Men of various trades, to the Sixteen Aldermen of Bristol* (Sixteen Working Men, 1871). The pamphlet decried the dearth of green space in the south and east of the city, stating that

> [Clifton and Durdham Downs] are mainly for rich people who can afford to live in that neighbourhood: it would take us an hour's walking, after the hard toil of the day is over, to get to these beautiful spots, and then another hour to get home, thus making pleasure a toil.
>
> (4)

Of particular interest is the manner in which the pamphlet seeks to make its point by expressing this pleasure in the language of genteel enjoyment and aesthetic refinement. It describes the flowers one would hope to find in such a 'people's park', given that 'at present we see little better than daisies and dandelions', continuing to express the desire to 'feel the grass under our feet, or sit with our wives on a summer's eve and watch our children play' (Sixteen Working Men, 1871: 4–5).

Following the critiques of the park movement described above, such a document may be firmly placed within the *embourgeoisement* of the working classes occurring over the Victorian era. Such expressions represent no more than the power of the park as a moral geography: its success in shaping the unquestioning subjectivities that would forestall the rise of another Chartism or Reform League. What I wish to highlight is the importance of the gesture of appropriating these aesthetic pleasures *for* the working population of Bristol as a disruption of the notion that they were below or unworthy of such experiences. The document questions the assumption that the articulation, and appropriation, of aesthetics of the park was the preserve of the philanthropic middle classes. Perhaps the most prominent and long-lasting example of how the symbolic codes inscribed in parks were exceeded and reappropriated is the regularised association football's becoming the site for specifically working-class tribalisms (Bailey, 1987: 186).

As Marne concludes, 'in the final analysis it is "the people" who creatively produce public space through their use of it and the values they ascribe to it'. By citing these examples, my aim is to demonstrate how the imposition of bourgeois codes of conduct, encouraging and inciting individuals to temperance and calm, assumes a smooth process imposed, without creative resistance, upon working-class populations. Yet these latter were, in different ways, developing their own particular attachments to park spaces. The full history of how 'the people' appropriated green spaces, developing their own languages of aesthetic pleasure, is yet to be written. What Rancière highlights is the importance of writing it; to change the story written about the working class in which disruptive subjects were forced into middle-class moulds as part of a process of power; to change the narrative wherein disordered youths are forced into more manageable expressions of discipline and belonging. The political act is to draw attention to the contingency of these stories and to act on the presupposition that these distributions of aesthetic capacities, in the end, do not hold.

Continuing into the present, rather than approaching parks in terms of distributions of aesthetic capacities and their becoming the site for certain techniques of power, we might look instead to the moments in which new languages of appreciation, experience and ownership are formed by young people. Rather than taking the determining spatial image to be the brightly lit and passively surveilled[4] Multi-Use Games Area, we might take instead the skate park, with its complex frameworks for acceptable graffiti and the interaction of different vehicles. More controversially, we might take the image of the ecstatic carnival of driving stolen vehicles. Jeremy Brent, whose text *Searching for Community; Power and Representation on an Urban Estate* (2009) details his experience as a youth worker and the tension between his understanding of, but ultimate distance from, the young people with whom he worked, describes his surveying a scene, in a local wooded area, of burned-out cars and motorbikes:

> Going around the area in the day (at night I would have been out of time and out of place), I could get a sense of the excitement that the young people must have felt, as well as seeing the destruction caused. It was like a scene after a carnival. I could almost hear the shrieks of delight from the evening before.
>
> (160)

Brent's account is careful not to romanticise this aesthetic life. 'These are not sustainable communities' (163), he states, noting that even without the intervention of the police the ecstatic nature of the experience precluded any enduring community culture. Yet he notes also the enjoyment and investment in the disciplinary aspects of these performances – in other words, the levels of order and learning amidst these seemingly disordered and destructive gatherings. Thus, instead of determining the park and its practices as an aesthetic field to be interpreted, we might turn to the manner in which Brent, as an observer, recognises the equality of the young people's uses of the space, the manner in which he experiences also the exhilaration, excitement and productivity of the illicit practices taking place there.

This leads us to a rather more speculative proposal on the role of the park within the city, namely, the idea that parks are uniquely *experimental* spaces. Within an increasingly segmented urban environment, green space provides a platform for new experiences, conjugations of aesthetic pleasure and disciplinarities. The critical narratives described in this chapter have accounted very well for the introduction of nature into the city as a carefully managed balance of control and disorder whose goal is the containment of disruptive and unruly bodies. I am proposing a new narrative: the continual emergence, at the seams of the park space, of nature as experiment, contingency and play. Despite continual attempts to control and manage green space, the continual return of communal illicit experiences – and of new ways of sharing and expressing these experiences – suggests that the park has been a particularly rich site for that disruption of a certain story in which all conform to their determinate aesthetic capabilities.

It is in this light that the forms of protest taking place through and in urban parks and other public spaces gain a new significance. Indicative of this were the June 2013 protests in Turkey, beginning (and returning) as they did from an 'occupy style' peaceful resistance to the development of a shopping mall on Gezi Park – 'a historically public park, an urban commons' (Kuymulu, 2013). As Kaya Genç, whose account of this time begins from the dissonance with his own, more placid and ordered experience of strolling through the park on his way to work, observed of the transition to a dynamic of creative resistance and violent police reaction:

> When I walked home that evening, dozens of tents had been pitched in Gezi. The night was warm. Bands played: celebrities gave readings; it was fun to booze in the face of the new ban on selling alcohol at night. . . . At dawn police attacked the protesters with tear gas and water cannon, then set their empty tents on fire.
>
> (Genc, 2013)

Accounts of the shared experience within these protests (see Feigenbaum, McCurdy and Frenzel, 2013) highlight the possibilities for creativity and political openness they afford and how they exceeded the demands – banking reform, democratisation – from which they were conceived (Žižek, 2013). They highlight how these politics were intertwined with the forms of living afforded by the reimagining of these spaces, and how attempts to negotiate the social issues raised were seized upon to discredit the politics of those involved (see Holehouse, 2011). Twelve days after the original Gezi Park occupation, after the authoritarian response depicted above had provided the trigger for wider protests, Mehmet Kuymulu noted that no 'victory' had yet been secured by those involved, 'except one'; 'the uprisings that caught AKP government of-guard brought together an unlikely body of people from all walks of life for the first time in recent memory'.

I propose a reimagination of the park as a privileged site for the commons, and as such a site for rethinking the meaning of 'community'. By taking the park, as a space of experimentation and play, rather than the residential neighbourhood, as the ground of community we might open new avenues for incorporating community into urban planning and development that avoid the tendency towards techniques for the management and containment of the 'usual suspects'.

Conclusion: the park ecology and its aesthetic disruption

The urban park is, as Byrne and Wolch note, a 'socially mediated ecology with deep roots' (2009: 745), an aesthetic space composed of historically situated discourses, techniques, ideologies and experiences. Central to any understanding of this complex ecology is the historical development of a disciplinary system for controlling *nature*, both that of the appetitive pleasures of the humble subject and that of the wild geography of the unenclosed commons. The strictly drawn lines of hierarchy and exclusion, and the civilizing gaze of the paternal middle-class, remains deeply woven into the material fabric of the park (Firth, 2003; Marne, 2001). The park remains designed, organised and policed to direct gazes and judgements both between users and into the park from the adjacent houses (Taylor, 1995; Lawrence, 1993).

Central to such arrangements of power, as Millie notes, is the cleansing of spaces of disorder. Anti-social behaviours have always taken place in even the most regulated parks (Marne, 2001: 436; Conway, 1991: 187), whose arrangement has from the outset been designed either to keep such behaviour out or to exert a 'civilizing' influence upon it. It may be observed that the 'community' inscribed in the park space is an exclusionary community; it is the orderly, responsible and morally virtuous community collectively enjoying the peace and fresh air (Byrne and Wolch, 2009: 747).

The purpose of this chapter has been to highlight a *different* aesthetics of green space. It has been to highlight not only the aesthetic techniques of ordering and governance, but the aesthetic in its disruption of hierarchies and orders: the aesthetic as a suspension of the distribution of the sensible policed by these techniques of governance. It has been to highlight the manner in which 'community' is inscribed in green space not only as the exclusionary community defined by the enforcement of a morally virtuous subjectivity, but as the commons: a constitutive opening to practices of aesthetic *dissensus*. In the final section I proposed that green space has played the role of presenting to the city this different commonality, one comprised not of a sharing within societal boundaries but of a suspension or disruption of these boundaries. Through an investigation of this supplementary tension between containment and opening, this chapter proposed that an engagement with these moments in the history of green space might open the sites for an enduring democratic politics of community. Against the accounts of a controlled, managed and governmentalised nature, the chapter has sought to present the park as the site of the 'uncertain community', in other words, as a figure within the city of how a disruption of the tightly policed roles that divide us might be achieved.

Underpinning these proposals is the Rancièrian presupposition of equality, which, I have demonstrated above, enacts a break with the Bourdieusian presumption that the mapping of tastes is *all there is* (and, as such, that the only possible political intervention lies in revealing to marginalised individuals their inferiority within the aesthetic system). On Rancière's account, the promise of transcendence in the aesthetic is directly critical; it is the practised and enacted presupposition of equality, one whose effect is the reconfiguration of structures of power. In short, the political intervention is *already* being enacted in multiple practices of aesthetic *dissensus*. Such a thought would reconfigure our understanding not only of aesthetics, but also of community; of what it is we share, what it is we have 'in common'. For returning to this concept of 'uncertain communities', what Rancière is describing, in stark contrast to the communitarian accounts that have dominated social scientific discussion on the subject, is the community as a *political event*: the disruptive act, or *dissensus*, of the 'uncertain community'. What Rancière asks, as a point of departure, is to be open to this aesthetic disruption, to be open to the commons as the illegitimate claiming of a part by the 'part which has no part' (1999: 30).

Notes

1 Save the few whose spaces were not affected by graffiti, all the group members interviewed considered a strict distinction between tagging, as a territorial practice, and the more artistic forms of graffiti. In the evolution of the discipline, however, these practices have always been firmly intertwined (Snyder, 2009).
2 In 2005 a renowned Banksy graffito in the city centre was allowed to stay after 93 per cent of respondents to an online poll on the city council website voted to keep it. The artist has been generally feted in the city, notably in an enormously popular exhibition in the city museum.
3 May is describing in this passage the 'demos', the term used in *Dissensus* (2010: 32–33) in an equivalent role to the 'uncertain community' here.
4 One technique to promote 'passive surveillance' is the placing of facilities within view of nearby houses.

Bibliography

Arat, Y. (2013) Violence, resistance, and Gezi Park, *International Journal of Middle East Studies*, 45(4): 807–809.

Armstrong, I. (2000) *The Radical Aesthetic*. Oxford: Blackwell.

Ashworth, A., Gardner, A., Morgan, R., Smith, A., Von Hirsch, R. and Wasik, M. (1998) Neighbouring on the oppressive, *Criminal Justice*, 16: 7–14.

Bailey, P. (1987) *Leisure and Class in Victorian England: Rational Recreation and the Contest for Control, 1830–1885*. London: Routledge & Kegan Paul.

Bird, W. (2004) *Natural Fit; Can Green Space and Biodiversity Increase Levels of Physical Activity?* London: Royal Society for the Protection of Birds.

Bourdieu, P. (1984) *Distinction: A Social Critique of the Judgement of Taste*. London: Routledge & Kegan Paul.

Brent, J. (2009) *Searching for Community: Representation, Power and Action on an Urban Estate*. Bristol: Policy.

Burney, E. (2005) *Making People Behave: Anti-Social Behaviour, Politics and Policy*. Cullompton: Willan.

Byrne, J. and Wolch, J. (2009) Nature, race, and parks: past research and future directions for geographic research, *Progress in Human Geography*, 33(6): 743–765.

Chakribarti, S. and Russell, J. (2008) ASBOmania, in Squires, P. (ed.), *ASBO Nation: The Criminalisation of Nuisance*. Bristol: Policy, pp. 307–318.

Conway, H. (1991) *People's Parks: The Design and Development of Victorian Parks in Britain*. Cambridge: Cambridge University Press.

Conway, H. (2000) Everyday landscapes: public parks from 1930 to 2000, *Garden History*, 28(1): 117–134.

Cranz, G. (1982) *The Politics of Park Design. A History of Urban Parks in America*. Cambridge, MA: Massachusetts Institute of Technology Press.

Dean, M. (1999) *Governmentality: Power and Rule in Modern Society*. London: Sage.

De la Fuente, E. (2000) Sociology and aesthetics, *European Journal of Social Theory*, 3: 235–237.

Deleuze, G. (1992) Postscript on the societies of control, *October*, 59 (Winter): 3–7.

Donajgrodzki, A.P. (1977) *Social Control in 19th Century Britain*. London: Croom Helm.

Feigenbaum, A., McCurdy, P., Frenzel, F. (2013) Towards a method for studying affect in (micro)politics: the campfire chats project and the Occupy Movement, *Parallax*, 19(2): 21–37.

Firth, A. (2003) State form, social order and the social sciences: urban space and politico-economic systems 1760–1850, *Journal of Historical Sociology*, 16(1): 54–79.

Gabriel, N. (2011) The work that parks do: towards an urban environmentality, *Social & Cultural Geography*, 12(2): 123–141.

Gibson-Graham, J-K. and Roelvink, G. (2010) An economic ethics for the anthropocene, *Antipode*, 41(S1): 320–346.

Goldsmith, C. (2008) Cameras, cops and contracts: what anti-social behaviour management feels like to young people, in Squires, P. (ed.), *ASBO Nation: The Criminalisation of Nuisance*. Bristol: Policy, pp. 223–238.

Griffin, E. (2005) *England's Revelry: a History of Popular Sports and Pastimes, 1660–1830*. Oxford: Oxford University Press.

Hubbard, P. (2004) Cleansing the metropolis: sex work and the politics of zero tolerance, *Urban Studies*, 41(9): 1687–1702.

Kuymulu, M.B. (2013) Reclaiming the right to the city: reflections on the urban uprisings in Turkey, *City: Analysis of Urban Trends, Culture, Theory, Policy, Action*, 17(3): 274–278.

Langman, L. (2013) Occupy: a new new social movement, *Current Sociology*, 61(4): 510–524.

Lawrence, H.W. (1993) The greening of the squares of London: transformation of urban landscapes and ideals, *Annals of the Association of American Geographers*, 83(1): 90–118.

Linebaugh, P. (2009) *The Magna Carta manifesto: liberties and commons for all*. Berkeley: University of California Press.

Marne, P. (2001) Whose public space was it anyway? Class, gender and ethnicity in the creation of the Sefton and Stanley Parks, Liverpool: 1858–1872, *Social & Cultural Geography*, 2(4): 421–443.

Matless, D. (1997) Moral geographies of English landscape, *Landscape Research*, 22(2): 141–155.

Millie, A. (2008a) Anti-social behaviour, behavioural expectations and an urban aesthetic, *British Journal of Criminology*, 48(3): 379–394.

Millie, A. (2008b) Anti-social behaviour in British cities, *Geography Compass*, 2(5): 1681–1696.

Moore, S. (2008) Street life, neighbourhood policing and 'the community', in Squires, P. (ed.), *ASBO Nation: The Criminalisation of Nuisance*. Bristol: Policy, pp. 179–202.

Olwig, K.R. (2005) Representation and alienation in the political land-scape, *Cultural Geographies*, 12(1): 19.

Prior, D. (2007) *Continuities and Discontinuities in Governing Anti-Social Behaviour*. Birmingham: University of Birmingham Press.

Prior, D., Farrow, K. and Paris, A. (2006) Beyond ASBOs? Evaluating the outcomes of anti-social behaviour initiatives – Early findings from a case study in one English city, *Local Government Studies*, 32(1): 3–17.

Rancière, J. (1991) *The Ignorant Schoolmaster: Five Lessons in Intellectual Emancipation*. Stanford, CA: Stanford University Press.

Rancière, J. (1999) *Disagreement: Politics and Philosophy*. Minneapolis: University of Minnesota Press.

Rancière, J. (2002) The aesthetic revolution and its outcomes. *New Left Review*, 14: 133–151.

Rancière, J. (2004) *The Politics of Aesthetics: The Distribution of the Sensible*. London: Continuum.

Rancière, J. (2006) Thinking between disciplines: an aesthetics of knowledge, *Parrhesia*, 1: 1–12.

Rancière, J. (2010) *Dissensus: On Politics and Aesthetics*. London: Continuum.

Roberts, J.M. (2001) Spatial governance and working class public spheres: the case of a Chartist demonstration at Hyde Park, *Journal of Historical Sociology*, 14(3): 308–336.

Rose, N. (1999) *Powers of Freedom: Reframing Political Thought*. Cambridge: Cambridge University Press.

Rosenzweig, R. and Blackmar, E. (1992) *A History of Central Park*. New York: Cornell University.

Sagar, T. (2007) Tackling on-street sex work, *Criminology and Criminal Justice*, 7(2): 153–168.

Scoular, J., Pitcher, J., Campbell, R., Hubbard, P. and O'Neill, M. (2007) What's anti-social about sex work? The changing representation of prostitution's incivility, *Community Safety Journal*, 6(1): 11–17.

Sennett, R. (1992) *The Uses of Disorder: Personal Identity and City Life*. London: W.W. Norton

Sixteen Working Men (1871) *The Cry of the Poor: Being a Letter from Sixteen Working Men of Bristol to the Sixteen Aldermen of the City.* Bristol: W. and F. Morgan.

Slaney, R.A. (1833) Report from the Select Committee on Public Walks: with the minutes of evidence taken before them. *House of Commons Papers*, Vol. XV (p. 337).

Snyder, G. (2009) *Graffiti Lives: Beyond the Tag in New York's Urban Underground.* New York: New York University Press.

Squires, P. (2008) The politics of anti-social behaviour, *British Politics*, 3(3): 300–323.

Stedman-Jones, G. (1971) *Outcast London.* Oxford: Clarendon Press.

Taylor, A. (1995) Commons stealers, land-grabbers and jerry builders: space, popular radicalism and the politics of public access in London, 1848–1880. *International Review of Social History*, 40: 383–408.

Trinder, B. (2000) Industrialising towns 1700–1840. In Clark, P. (ed.), *The Cambridge Urban History of Britain, Volume 2.* Cambridge: Cambridge University Press, pp. 805–829.

Tufekci, Z. (2013) The medium and the movement: digital tools, social movement politics, and the end of the free rider problem, *Policy and Internet*, 6(2): 202–208.

Wheater, C.P., Potts, E., Shaw, E.M., Perkins, C., Smith, H., Casstles, H., Cook, P.A., Bellis, M.A. (2007) *Urban Parks and Public Health: Exploiting a Resource for Healthy Minds and Bodies.* Manchester: Department of Environmental and Geographical Sciences, Manchester Metropolitan University Press.

Wylie, J. (2002) An essay on ascending Glastonbury Tor. *Geoforum*, 33(4): 441–454.

Young, C. (1998) The making of Bristol's Victorian parks. *Transactions of the Bristol and Gloucestershire Archaeological Society*, 116: 175–185.

Young, T. (2004) *Building San Francisco's Parks, 1850–1930.* Baltimore, MD: Johns Hopkins University Press.

Žižek, S. (2012) Occupy Wall Street: what is to be done next? *Guardian*, 24 April. Retrieved from http://www.guardian.co.uk/commentisfree/cifamerica/2012/apr/24/occupy-wall-street-what-is-to-be-done-next

Žižek, S. (2013) Trouble in paradise: the global protest. *London Review of Books*, 35(14): 11–12.

Index

Milton Keynes UK
Ingram Content Group UK Ltd.
UKHW040104071024
449327UK00019B/798